AF344342

PULSE & SWITCHING CIRCUITS

By Harvey F. Swearer

Robert E. Krieger Publishing Company
Malabar, Florida
1984

Original Edition 1970
Reprint Edition 1984

Printed and Published by
ROBERT E. KRIEGER PUBLISHING COMPANY, INC.
KRIEGER DRIVE
MALABAR, FLORIDA 32950

Copyright © 1970 by TAB BOOKS
Reprinted by Arrangement

Printed in the United States of America

Library of Congress Cataloging in Publication Data

Swearer, Harvey F.
    Pulse & switching circuits.

    Reprint. Originally published: 1st ed. Blue Ridge Summit, Pa. :
Tab Books, c1970.
    Includes index.
    1. Pulse circuits.   2. Switching circuits.   I. Title.
II. Title: Pulse and switching circuits.
TK7868.P8S94   1984   621.3815'34   84-904
ISBN 0-89874-739-2

Preface

For anyone interested in the science of electronics, and particularly those involved in its practical day-to-day aspects, the use of pulse and switching circuits is a vital concern. There is hardly a phase or field of electronics untouched by some form of pulsing and switching applications. Therefore, it is in his own best interest that every individual develop a familiarity with the subject, lest he find himself completely outdated.

This is not an engineering thesis; it is a practical examination of pulses, the circuits that shape them, and the variety of ways they can be put to work. We begin with an examination of pulses, basic definitions, and pulse parameters. From there we logically progress to pulse generators and circuit response characteristics.

After establishing this firm foundation, we devoted the remainder of the book to a variety of practical, everyday applications: remote control, TV, computers, radar, telemetry, and various automation devices, so that regardless of your niche in the electronics world you would find something of a useful nature.

Harvey   F.   Swearer

Contents

Chapter 1

Pulses

The pulse is ever on the rise (no pun intended) in the field of electronics, and as we continue to find new uses for this nonsinusoidal wave, the importance of a clearer comprehension becomes increasingly more apparent to the technician-engineer. An attempt is made to at least cover the more popular uses of pulse circuitry—automation, digital computers, electronic accounting, food dispenser, instrumentation, measuring devices, remote-control, (wired and wireless), telemetry, telephony, traffic control, space communications, and sensing units. Those currently in greatest use in our ecomony are covered extensively, and others of lesser performance only lightly as they intersect the main topic of discussion.

One of the first things that must be understood regarding pulse circuitry is that there is a vast difference in the way electronic circuits treat pulse waveforms. Since most pulses are basically rectangular in form instead of sinusoidal, they are subject to a variety of influences considered inconsequential in sine-wave circuits. Even the capacitance and inductance ever present in the very leads or wiring of most circuits present problems in handling pulse waveforms; therefore, the rectangular pulse requires frequent corrective measures to eliminate this built-in distorting characteristic. In order to emphasize the effect of capacity and inductance on a pulse, we must simply recall the fact that XL and XC formulas are useless when dealing with these waves.

As stated, most pulse-circuit uses require rectangular waveshapes, which are highly susceptible to all kinds of distortion. We must remember that the shape of the common sine wave is altered only by nonlinear devices (tubes, transistors, rectifiers), but the rectangular pulse is badly distorted in shape by nonlinear devices **and** L or C elements as well. Rectangular pulse waves are rich in harmonics—the sharp corners contain mainly high-order harmonics. Since rec-

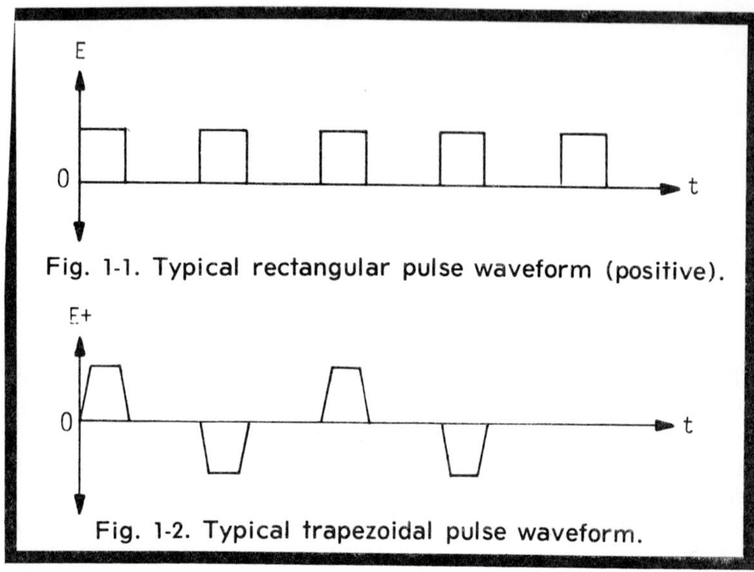

Fig. 1-1. Typical rectangular pulse waveform (positive).

Fig. 1-2. Typical trapezoidal pulse waveform.

tangular waves contain so many harmonics of sufficient amplitude to require consideration, they are most difficult to analyze by individual components. Harmonic analysis, although perfectly reliable, is much too time consuming to be useful to the technician-engineer.

BASIC DEFINITIONS

The nonsinusoidal wave may be typed according to its "occurrence," i.e., aperiodic or random interval, as in digital computers and static disturbance and periodic, those waves occurring at regular intervals as a square or sawtooth wave. Fundamentally, a pulse is a wave of constant value and definite length that varies in rise time and decay (fall) time. It may also be stated as a rise and fall of voltage of relatively short duration. This nonsinusoidal waveform of great interest and widespread use in electronics today is usually almost entirely confined to one side of the zero-voltage axis, thus giving it a DC component. An ideal rectangular pulse wave is shown in Fig. 1-1 and a trapezoidal pulse wave in Fig. 1-2. The pulses in Fig. 1-1 are periodic (occur at regular intervals in time), and since they have perfectly straight sides and sharp corners they are ideal pulses instead of actual.

Electronic generators of the magnitude necessary to produce the infinite number of harmonics of the perfect pulse are non-existent. In actual practice, the circuit components can accommodate only harmonics lying within the upper and lower limits of their frequency response capability, and so the true wave has rounded corners and sides, too. Fig. 1-3 displays exactly what is meant, but normally the differences appear less noticeable to some degree. There is a definite similarity between actual rectangular and trapezoidal pulses, due to the limited high-frequency response of just about any type of electronic equipment. The pulse voltage is unable to rise instantly from zero to maximum as the straight side of the leading edge in the ideal wave would require. The practical pulse can never have truly vertical sides, but the closer we come, the higher the frequency response of our pulse-handling circuitry must be in order to avoid distorting that pulse.

PART NOMENCLATURE

Rise time—elapsed time required for a pulse to move from 10 percent up to 90 percent of its maximum value.

Decay time—elapsed time required for a pulse to fall from 90 percent down to 10 percent of its maximum value. Also known as fall time.

Duration—time between the end of the rise time and the beginning of the decay time.

Width—sum of rise, decay, and duration times.

Rest time—time between successive pulses, which may vary widely from pulse to pulse.

PRT—pulse repetition time or pulse repetition period is the time between a specific point on one pulse and the corresponding point on the adjacent pulse.

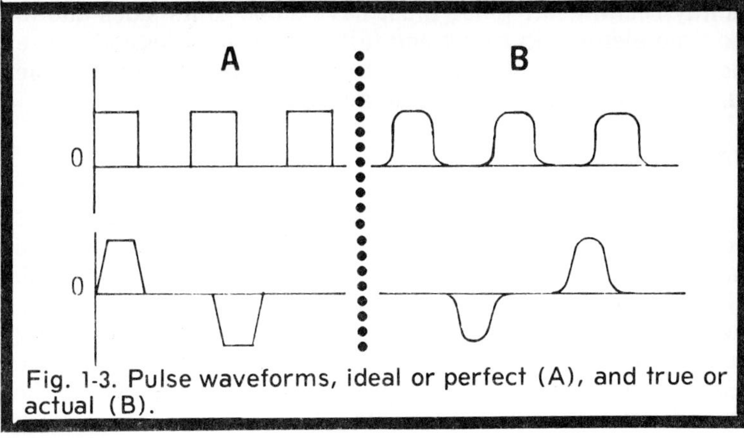

Fig. 1-3. Pulse waveforms, ideal or perfect (A), and true or actual (B).

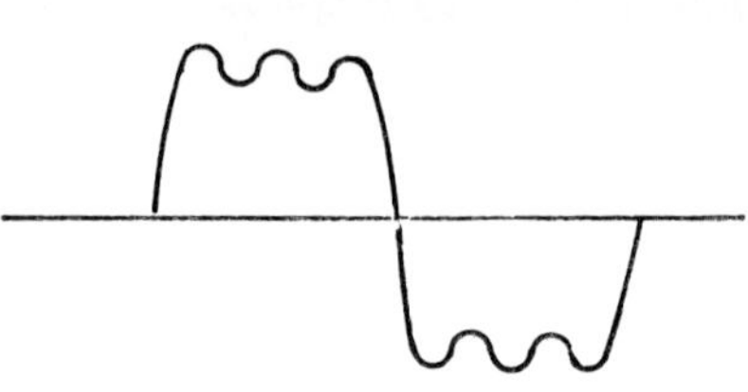

Fig. 1-4. Waveform of a sine wave with 3rd and 5th harmonics added.

PRF–pulse repetition frequency or pulse repetition rate is the average number of pulses that occur per second.

Amplitude–indicates the magnitude (or "strength") of the pulse. (A complete list of pulse definitions appears at the end of this chapter.)

Pulses are not restricted as to shape, amplitude, or duration, but–unlike sine waves–they lack continuity. The terms pulse repetition time and pulse repetition frequency should not be confused, although one is the reciprocal of the other, i.e., PRT equals 1/PRF and PRF equals 1/PRT. As an example, assuming a pulse repetition time of 5 microseconds the pulse repetition frequency would equal 1/5 microseconds or $1/5 \times 10^{-6}$ equals 200,000 pulses per second. In other words, the shorter the pulse repetition time, the higher the frequency, and likewise the higher the capability of the pulse-handling equipment before distortion slips in, changing the waveshape. As any rectangular pulse must have slope to its sides due to the time element in rising and falling, it is only logical to use the 10 and 90 percent figures to facilitate measurement of the duration time with reasonable accuracy.

HARMONICS

A whole-number relationship between two sine waves would indicate harmonics. If the fundamental is 1 Hz, the third harmonic would be three Hz; by the same token, the tenth harmonic would complete ten Hz in that same time interval. Sometimes harmonics are identified as odd or even, referring to the number with respect to the fundamental–third, fifth, seventh, ninth, etc., in the odd group, and in the even group the second, fourth, sixth, eighth, etc.

10

If a waveform is produced by a resonant circuit, it must be composed primarily of but one frequency, since a resonant circuit is designed to respond to a single, pure sine wave. However, in most other instances, the actual signal consists of many frequencies combined. The human voice is definitely not a single frequency, but rather a combination of many frequencies from a low of about 100 Hertz to approximately 3000 Hertz at the high end. If a series of electrical impulses representing the human voice were applied to a tank circuit, the circuit would respond to those impulses equaling its own resonant frequency. Should the resonant frequency fall between 100 Hz and 3000 Hz, the tank would emphasize that particular frequency. As may be expected, the output of the tank would be a sine wave.

No matter how complex, and regardless of shape, all waveforms are merely combinations of sine waves. Fig. 1-4 shows the appearance of adding the third and fifth harmonics to the fundamental, and that the rough resemblance to the square wave shape is very noticeable. If suitable proportions of the fundamental and all odd harmonics are combined, the total is a true square wave as displayed in Fig. 1-5. Hence, a square wave is not just a single frequency like the sine wave, but consists of many frequencies including the fundamental and all odd harmonics thereof. It is true that any waveshape that fails to meet the definition of a sine wave can be proven to be made up of a fundamental sine wave and certain harmonic sine-wave frequencies of that fundamental.

If an amplifier distorts a sine wave, the wave loses its sine-wave shape; therefore, it must be composed of more than one frequency, such as a fundamental and certain harmonics. For this reason, amplitude distortion in an amplifier is joined by the generation of spurious or unwanted harmonic frequencies that were not induced into the input of that amplifier. Fig. 1-6 shows the fundamental and the resultant waveform when the second, third, fourth, and fifth harmonics have been combined. Since the waveform has the general appearance of the teeth in a saw, it realistically bears the

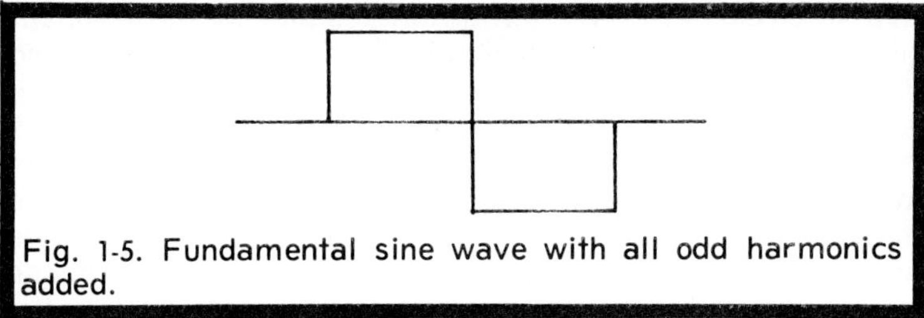

Fig. 1-5. Fundamental sine wave with all odd harmonics added.

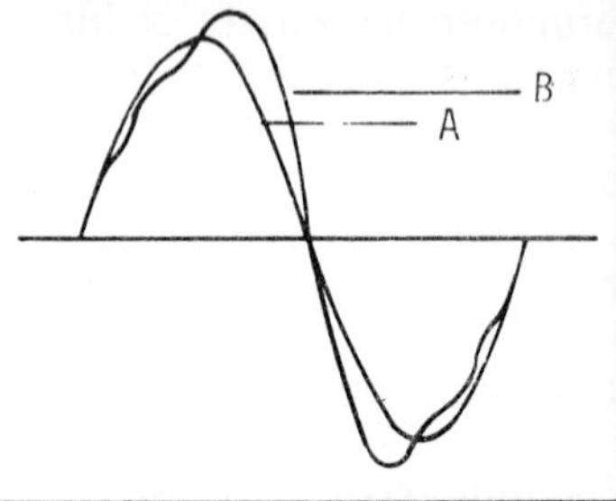

Fig. 1-6.   A fundamental sine wave (A) with the composite of the fundamental, plus 2nd, 3rd, 4th, and 5th harmonics superimposed (B).

name, sawtooth wave, and is considered to be in perfect form when the fundamental and all harmonics are combined. The square wave in Fig. 1-5 is perfectly balanced–the bottom half and the top half being mirror images. This is not so in the case of the sawtooth wave in fig. 1-6; even though the same amount of wave exists above and below the zero line, the two parts are not symmetrical. If the two halves of waveform B in Fig. 1-6 were a mirror image of each other, the very slight curve would appear at the start rather than the finish of the negative portion. In order to be perfectly symmetrical, the waveform can contain odd harmonics only; if it contains any even-order harmonics, it cannot be symmetrical regardless of what other harmonics, if any, are present.

RESONANCE

The characteristics of series and parallel circuits are opposite, of course. The parallel circuit impedance is high at resonance, and as the frequency departs from resonance the impedance rapidly decreases to a low value. The series circuit has a very low impedance at resonance, and as the frequency leaves resonance the impedance rapidly increases to a high value. How high the impedance at parallel resonance, or how low at series resonance, depends entirely upon the ohmic resistance in the resonant circuit. If that resistance is zero, the series circuit will have zero impedance at resonance and the parallel circuit will have infinite impedance at the resonant point. Needless to say, practical circuits do not have zero resistance, since there is always the resistance of the wire used in the inductor. This resistance should be kept as low as possible, because the lower the resistance, the lower the impedance in a series resonant circuit and the higher the impedance in a parallel resonant circuit. Low coil resistance means a high Q, and the higher the Q of the coil the lower the impedance at series resonance and the higher at parallel resonance. The Q is actually a figure of merit: Q equals XL divided by R and the higher the Q, the narrower the bandwidth as indicated in the formula:

$$Q \text{ equals } \frac{fo}{bw}$$

Resonant circuits have proven useful in electronics because of their high impedance at some frequencies and low impedance at others. In this way, they differ from resistors which have a comparatively high impedance at all frequencies. Resonant circuits are used in place of resistors where it is desired to have an appreciably higher or lower impedance at one selected frequency.

Impedance is the total opposition to current flow and is made up of resistance and reactance. In a DC circuit where there is no reactance, the impedance is equal to the resistance. In a circuit without resistance, like a capacitor across an AC voltage, the impedance is equal to the reactance. Reactance is that opposition to current flow which tends to cause a 90-degree phase shift between current and voltage. If a circuit has both reactance and resistance, the reactance tends to shift the current 90 degrees and the resistance tends to keep the voltage and current in phase. So, we have a compromise in which the current is less than 90 degrees out of phase from the voltage. The higher the resistance and the lower the reactance, the lower the degree of phase shift.

Line current is in phase with the line voltage when the parallel or series circuit is at resonance. At this time, the reactance of a series or parallel circuit at resonance is zero because the inductive and capacitive reactances cancel each other. If the circuit had any reactance whatsoever, the current would lead or lag the voltage.

PEAK VALUE

Fig. 1-7 shows a few examples of AC waves that are not in the sine-wave category. As these waves are made up of repeating patterns, terms like period and frequency, as well as formulas relating thereto, are definitely applicable. Each repeating pattern represents one Hz of the wave and the length of one Hz is a period. Therefore, the number of Hz per second is the frequency. The sawtooth wave (Fig. 1-7A) has a peak value of 50 volts and the square wave (Fig. 1-7B) has a peak of 30 volts. The peak-to-peak value is the total value of the wave swing from highest to lowest, and since the waves rise positively to the same level as they fall negatively, their peak-

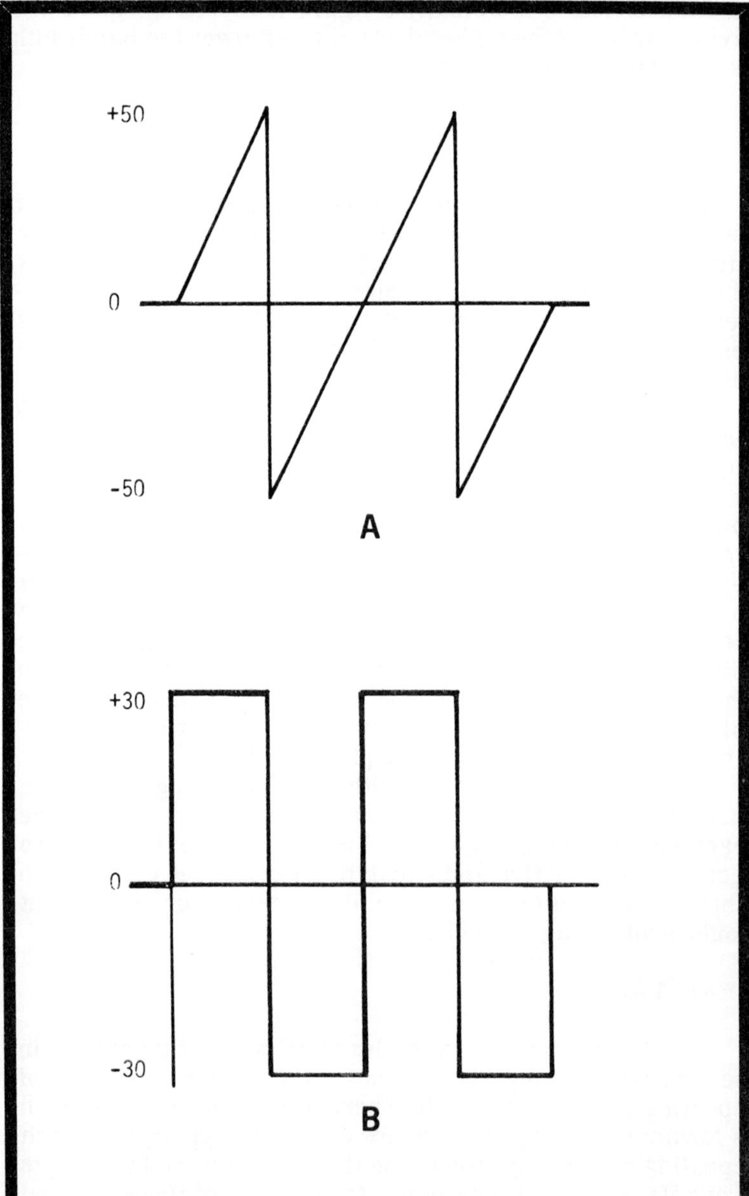

Fig. 1-7. AC waveforms (both positive and negative components) that are not sine waves.

to-peak value is exactly twice the peak value, or the peak value is one-half the peak-to-peak reading. Since the sawtooth varies from -50 volts to +50 volts, the peak-to-peak value is 100 volts.

PRIMARY FREQUENCY STANDARD

A compact, self-contained primary standard of the atomic-beam type, utilizes a cesium-beam resonator for stabilization of a high-quality quartz oscillator, resulting in an accuracy of plus or minus 1×10^{-11}. The control circuit provides continuous monitoring of the output and automatic logic circuitry presents an indication of correct operation. The cesium-beam frequency standard with clock and digital divider and automatic synchronization is pictured in Fig. 1-8.

TRANSISTOR PULSE CHARACTERISTICS

Transistor switching speed is the ability of a semiconductor to conduct and cut off in rapid succession; i.e., switching speed is determined by its transient characteristics. Therefore, the transient characteristics relate to the cut-off frequency, which, in turn, governs the switching speed. The steady-state and transient characteristics are the most important areas for consideration by the designer: The first provides the discrete levels to be maintained while the latter stipulates the speed of operation.

A transistor's nonlinear operating regions are determined by the steady-state characteristics, and when properly biased a large signal pulse causes it to function in the nonlinear region. A rectangular input drives the transistor from cut-off to saturation and then back to cut-off, thus operating in the nonlinear area and resulting in an output that differs greatly from the input. A voltage that undergoes an instantaneous change in amplitude from one level to another is a unit step voltage and is formed when the voltage rises immediately in the positive or negative direction.

In pulse and switching circuitry, the applied signal is normally of sufficient amplitude to cause the transistor to switch from on to off, or off to on, and could be referred to as the unit step voltage; see Fig. 1-10. The step voltage does not need to increase from zero to a positive level because the change from a negative voltage to zero still involves a positive-going step voltage. When the transistor is in the cut-off or saturation regions without an input signal, it is called the quiescent state. The active region is the only one where nor-

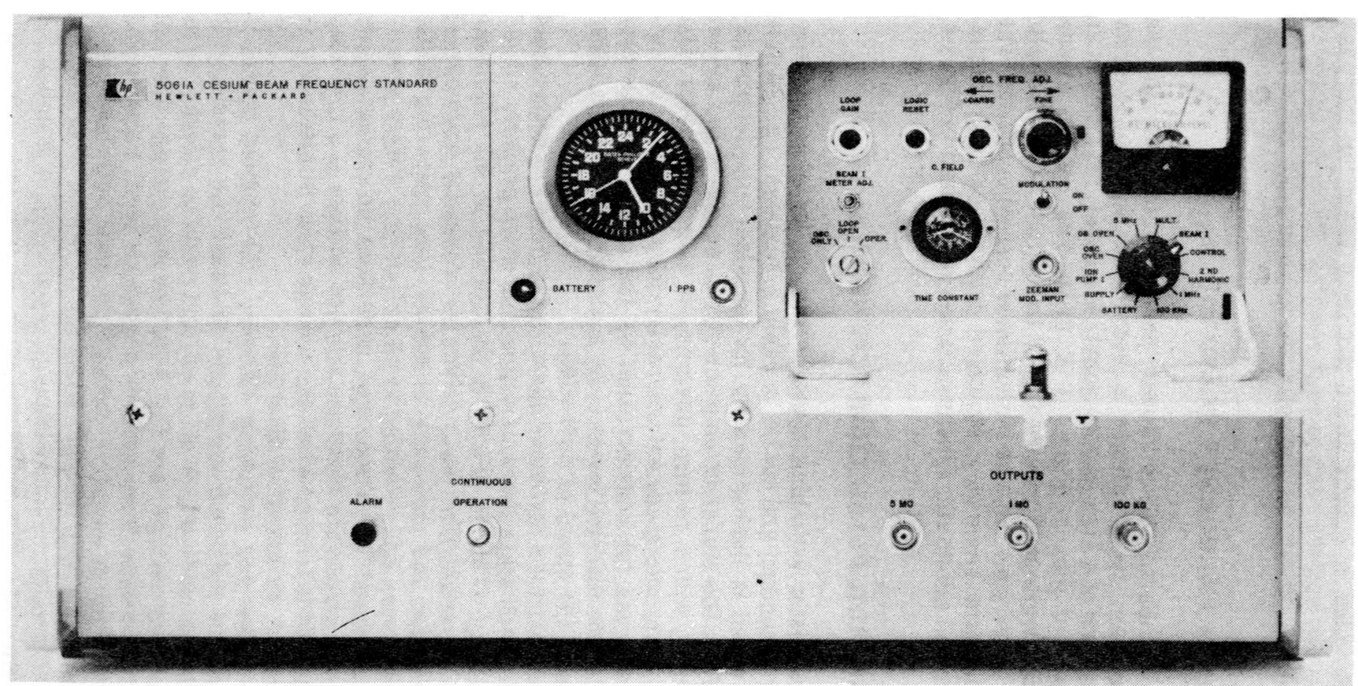

Fig. 1-8. Cesium beam frequency standard, Model 5061A, with a built-in clock and standby power supply. (Courtesy Hewlett-Packard)

mal amplification takes place, and in going through this
region the transistor reaches its limits of operation at
saturation or cut off. The transistor switch in operating from
off to on passes through the transient (active) region, and in
switching back from on to off again passes through the
transient region.

Normally, the common-base configuration offers better
transient characteristics than the common-emitter circuit,
which enhances its use in high-speed transistor switching
circuits; see Fig. 1-9. The PNP transistor acts as a switch for
positive pulses while the NPN provides switching action with
negative pulses when the common-base arrangement is offered. In the event the transistor is unable to respond instantaneously to any change in signal level, the output current
pulse will be distorted.

Distortion is noted in the four fundamental pulse
characteristics—namely, rise time, pulse duration, storage
time, and fall time. The slope of the leading edge of the output
pulse depends on the nonlinear characteristics and circuit
arrangement of the transistor. These same factors are also
responsible for the rise time, and a decrease in this charac-

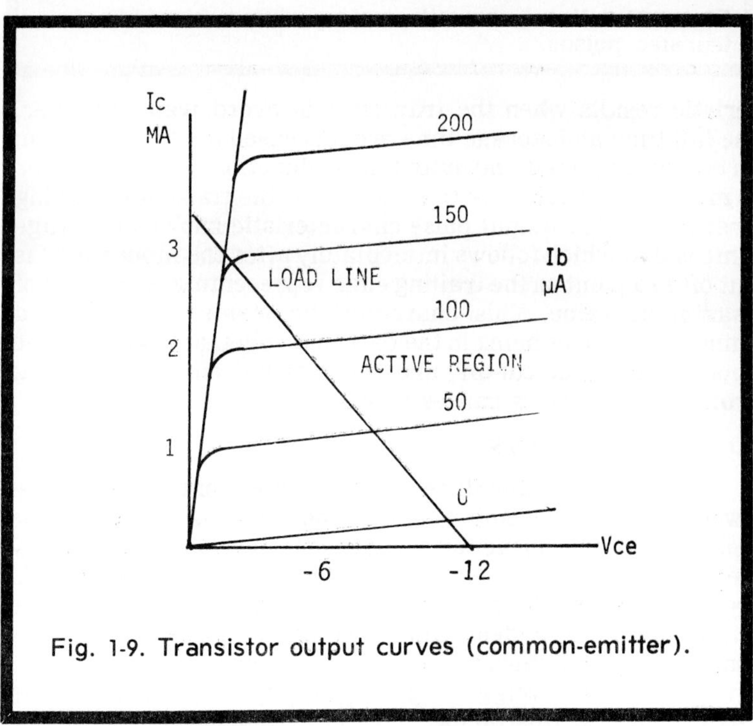

Fig. 1-9. Transistor output curves (common-emitter).

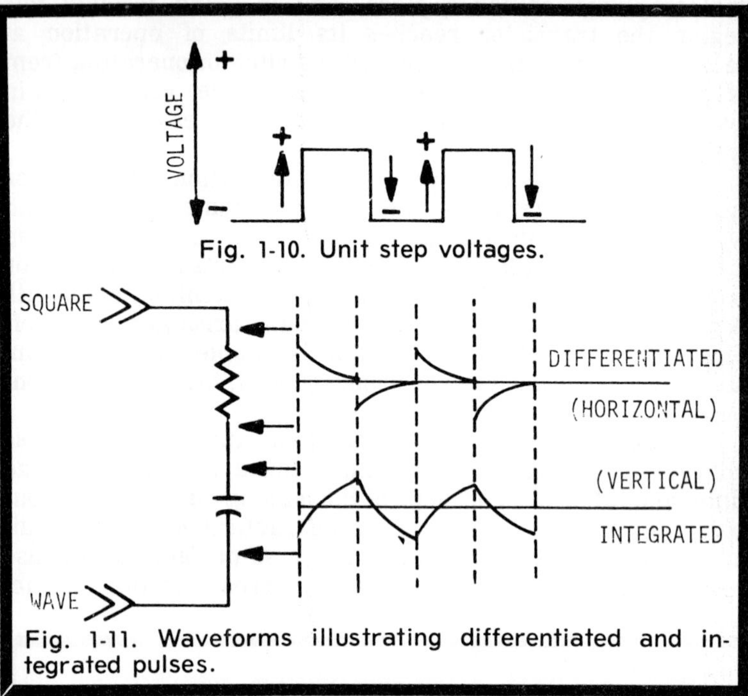

Fig. 1-10. Unit step voltages.

Fig. 1-11. Waveforms illustrating differentiated and integrated pulses.

teristic results when the transistor is overdriven. However, the fall time and storage time are increased by this operation. In comparing input and output pulse duration, when the latter is much longer, it is safe to assume that the transistor is being overdriven. The output pulse characteristic known as storage time is that which follows immediately after the input signal is cut off to a point on the trailing edge representing 90 percent of maximum value. This characteristic stems from injected minority carriers being in the base and collector regions at the moment of input cut-off, and the residual current resulting from these carriers causes storage time.

LOADING EFFECTS

One of the major differences, as mentioned before, between a sine wave and a pulse (nonsinusoidal) wave is the continuity of the former as opposed to the complete lack of any regular sequence in the latter. The spaces or rest periods between pulses are not part of the signal and may even be called off time or dead time. The steepness of the pulse is undoubtedly the most important parameter, and the timing, switching, or triggering capabiiity emphasizes the significance of this factor.

18

The proof of the importance of timing ability can be demonstrated by the action of TV sync pulses in locking the vertical and horizontal sweep systems of a TV receiver to that of the transmitter. The fact that the vertical and horizontal pulses differ only in width (although providing functions far apart) requires unusual techniques to separate the two signals from the composite video. Differentiating and integrating networks produce these new waves from the applied square wave. The RC circuit and response curves are shown in Fig. 1-11. The current from the square wave results in a differentiated pulse across the resistor (caused by the capacitor-charging current flowing through it) and an integrated pulse across the capacitor (due to the charge-discharge rate of the capacitor). The narrow, steep pulse is especially vulnerable to loading by meters or amplifier inputs and the amplitude would be reduced to such a degree as to render the pulse completely useless. The horizontal sync pulse has an extremely short time constant and is actually almost vertical in shape. It may be assumed that any measurements with meter or scope would reduce the level and alter the shape to make it meaningless.

PULSE FRONT STEEPNESS

Many analyses of pulsed circuits assume that the wavefront of a pulse is a step function, as at Point A in Fig. 1-12. Actually, it is impossible for a voltage to change instantaneously from one value to another. Therefore, the wavefront cannot have infinite slope as implied by this function. There certainly are cases where this assumption of perfectly square pulse fronts leads to erroneous results. In

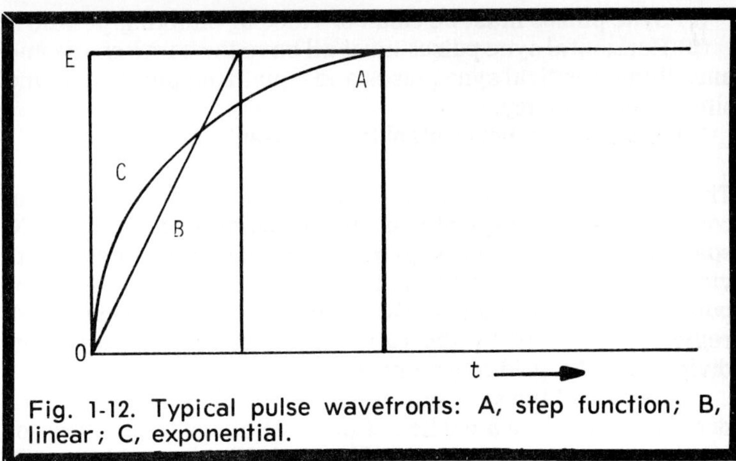

Fig. 1-12. Typical pulse wavefronts: A, step function; B, linear; C, exponential.

such cases the following analysis gives superior results with no greater complexity.

It is often appropriate to assume an infinite slope as a first approximation and to interpret the results accordingly. There are times when this gives incorrect results (an analysis of pulse differentiation), but a second approximation, a linearly rising pulse front (B in Fig. 1-12) which is close enough to be of practical help, can be assumed. A fairly complete analysis of the problem assumes an exponential rise of the pulse as at Point C in Fig. 1-12—a third and even closer approximation. A linear rather than exponential rise may at times be a more accurate assumption, as the rise is often linear until just before flattening out, particularly when peaking circuits are used.

The practical designer of electronic equipment prefers a few rules of thumb for quick application to various cases, especially as numerous stray effects need to be weighed and the circuit values are not accurately known.

(1) If the time constant is much greater than the rise time of the pulse, the pulse front is transferred across the circuit with but a slight loss in voltage.

(2) If the time constant of the circuit is about equal to the rise time of the pulse, the voltage output maximum is about two-thirds of the input.

SYNC PULSE SEPARATOR ANALYSIS

The ideal sync separation circuit requires:

(1) Sync pulses must be clean of video or blanking pedestal.

(2) Horizontal sync pulses must all have the same shape and amplitude. Vertical sync pulses and equalizing pulses demand similar consistency.

(3) Reasonable noise blanking is essential.

These limitations hold true for all levels of modulation; of course, the percentage of sync and video must be within FCC specifications. The sync separator input signal is a composite video with positive sync and the output presents the sync pulses only. The point of DC restoration in the sync pulse region is governed by the ratio: grid-input series resistance divided by the grid-to-ground R.

The tops of the sync pulses are removed as grid current is drawn, attenuating a portion of the sync above the restoration level by the ratio: cathode-grid R divided by the grid-input R.

The bottoms of the sync pulses and video are eliminated because they are below the low cut-off point of the tube or transistor. Since the grid "restores" near the blanking level, noise immunity is satisfactory, although the level is more critical at this point than at the sync tips. This is due to the normal variation in average video or the advent of the vertical sync pulse level causing a change in shape or width of the sync pulses at the output of the sync clipper, since the sync pulse is trapezoidal and not rectangular in shape. Variations due to changes in the average video signal are usually slow enough to be insignificant. The effect of vertical sync can be controlled by using a larger coupling capacitor. The only restriction on the capacitor value is the noise factor; some types of noise may be introduced by too large a value.

NOMENCLATURE AND DEFINITIONS RELATING TO PULSES

Bandwidth: Smallest continuous-frequency interval beyond which the amplitude of the spectrum does not exceed a prescribed fraction of the amplitude at the specific frequency.

Carrier: Pulse train used as carrier.

Delay unit: Unit with long time constant for delaying the arrival time of pulse.

Demoder: Constant delay discriminator.

Discriminator: Device which responds only to a pulse having a particular characteristic such as duration, period, amplitude, etc.

Doppler system: Pulsed radar system utilizing the Doppler effect to obtain information about a target.

Droop: Distortion of flat-topped rectangular pulse, described by a decline of the pulse top.

Duration, coder: Navigation device that generates coded transmissions by varying pulse lengths.

Duration, modulation: Pulse-time modulation in which the duration of pulses is varied.

Duty factor: Product of the average pulse duration and the pulse repetition rate.

Generator: Device or circuit for generating pulses.

Interleaving: Process for combining pulses from multiple sources in time-division multiplex for transmission over a common path.

Interrogation: Transponder triggering by a pulse or pulse mode.

Interval modulation: Pulse-time modulation in which pulse spacing is varied.

Jitter: Spacing variation in a pulse train due to random or systematic causes.

Mode multiplex: Device for selecting channels by pulse modes (multiple channels on the same carrier frequency).

Modulation: Carrier modulation by a pulse train.

Modulator: Device for applying pulses to the element where modulation occurs.

Position modulation: Pulse-time modulation where the position of pulse is varied in time.

Repeater: Device which receives pulses from one circuit and transmits the corresponding pulses into another.

Shaper: Transducer for changing the characteristics of pulse.

Spacing: Interval between the times of successive pulses, measured between the corresponding time or point on the waveform.

Spike: Spurious or unwanted pulse superimposed on the desired pulses.

Tilt: Distortion of the flat top on a rectangular pulse, described as a decline or rise of that pulse top.

Chapter 2

Pulse Generators

Basic pulse-forming circuits fall into three general groups according to function and purpose. The multivibrator is a two-stage amplifier with positive feedback, while the blocking oscillator, although similar to the "multi," requires only a single stage. Either the multivbrator or the blocking oscillator may be biased for free-running or triggered operation, but to obtain feedback the multivibrator uses RC coupling and the blocking oscillator utilizes inductive coupling. A third type of pulse-forming circuit is called a time-base generator and has the capability of producing a voltage or current that varies with time in some assigned way. The usual output of the time-base circuit is a sawtooth waveform.

MULTIVIBRATORS

Astable Multivibrator

The multivibrator (or "multi") using a grounded-emitter configuration bears one-shot, flip-flop, or free-running designations. The astable or free-running multi is used for generating square waves, timing frequencies, and frequency division. Fig. 2-1 is a practical astable multi circuit with the transistors DC biased so that both can conduct simultaneously except for the cross-coupling capacitors which prevent such action and insure alternate conduction of the transistors. T equals Ct + 100 divided by 28.8 us, with Ct in picofarads and us indicating microseconds. A synchronizing pulse would lock the multivibrator to an external oscillator fundamental or sub-harmonic as desired.

The circuit shown in Fig. 2-2 makes use of a unijunction transistor (UJT). When power is applied, the transistor is reverse-biased and C1 charges through D1 and R1. C1 can charge only to the maximum voltage drop across R2, and the C1 voltage must be capable of firing the transistor. As this

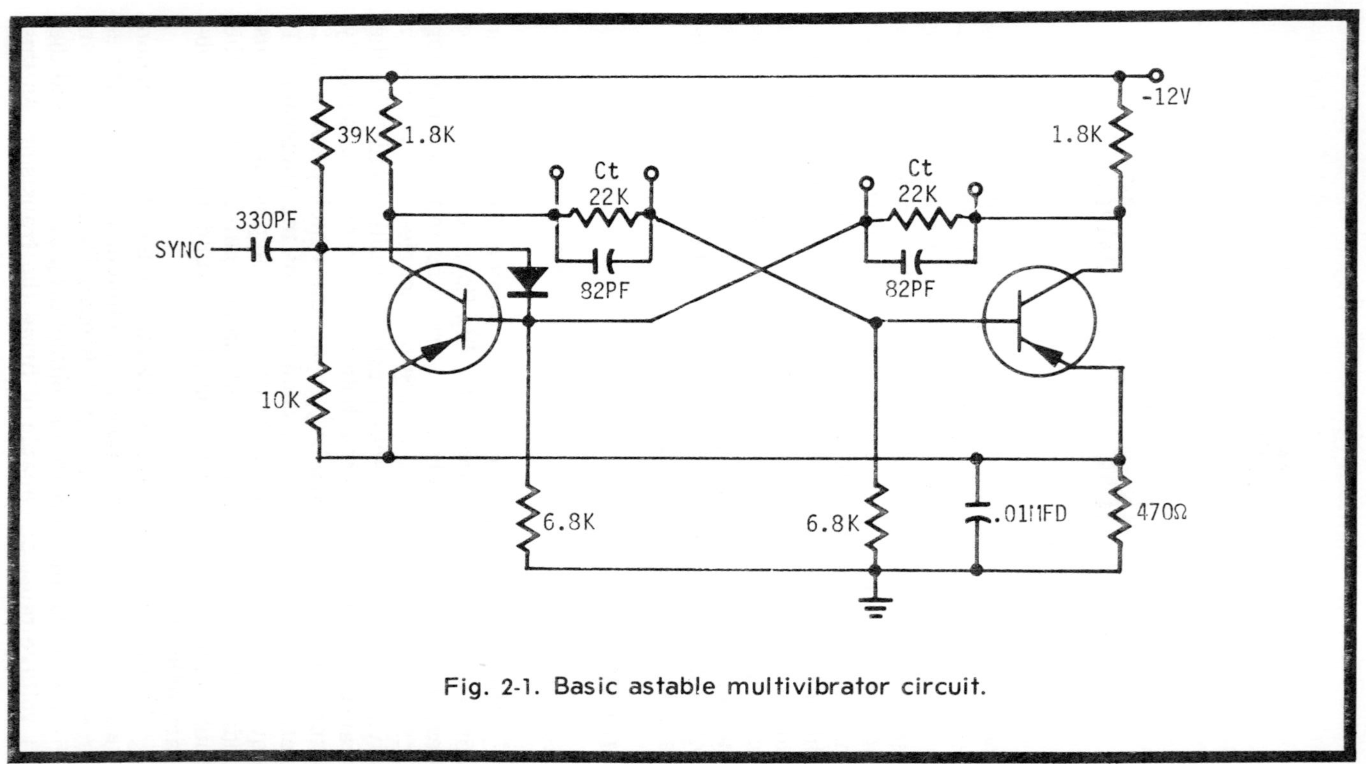

Fig. 2-1. Basic astable multivibrator circuit.

takes place, D1 is cut off and the UJT remains fired for the time constant provided by R2 x C1. At firing time, the emitter voltage drops near ground potential, permitting reverse-bias across the diode (cathode more positive than anode) which produces an open switch and isolation. During the diode switch open period, the transistor is still fired and current flows from negative through base 1, the PN junction, and R1 to positive. Now C1 is discharging through R2 , and the R2 voltage drop decreases slowly. As the cathode of D1 becomes more negative than the anode, the diode conducts and the switch is closed. The UJT is now reverse-biased and C1 recharges once more. The period when the diode switch is open determines the pulse width of the generator and is governed by the R2-C1 time constant.

Stability of the Astable Multi (Free-Running)

As a result of the frequency of the multivibrator being dependent on the resistance-capacitance time constant, it is natural that any change in these values as a result of temperature variation will cause the frequency of the circuit to change. In order to operate under variable temperature conditions, it is mandatory that the components used are

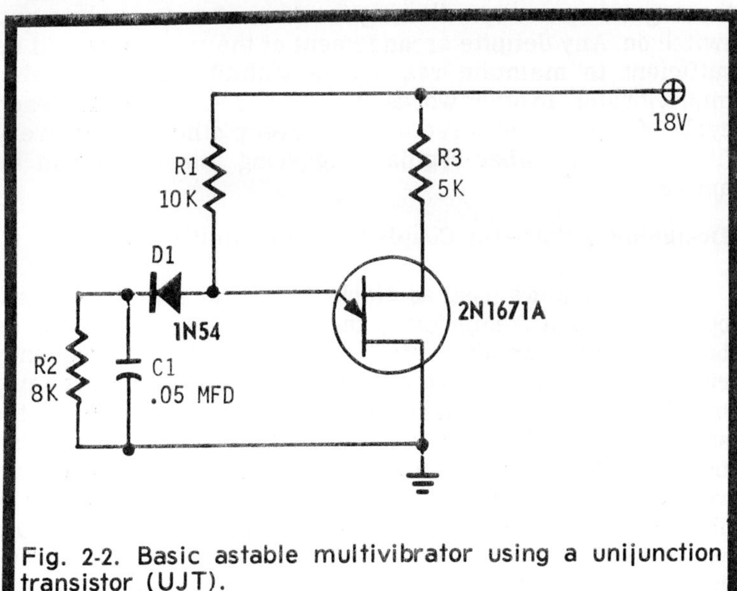

Fig. 2-2. Basic astable multivibrator using a unijunction transistor (UJT).

capable of maintaining specific values under adverse temperature conditions. Although the overall change as a result of these conditions may be reduced to comparative insignificance, the Icbo increase with temperature cannot be avoided.

The timing capacitor discharges through the base-biasing resistor and controls the frequency, but in so doing also furnishes the leakage current, Icbo, which varies considerably with temperature. As the discharge time of the capacitor is directly affected, frequency changes as a result may be held to a minimum by the use of a large capacitor and a low value base-biasing resistor. By selecting a base-biasing resistor of much lower value than the base-collector reverse-bias resistance, the leakage current will amount to a mere fraction of the discharging current of the capacitor and frequency variations will be slight.

In the event of extremely accurate frequency requirements, a multivibrator may be locked in step by a synchronizing pulse, originating in a crystal-controlled oscillator. By applying sync pulses to the off transistor base (as the sum of the sync and bias voltages reaches a negative value), the transistor switches on. During the time the transistor is off, the base is slightly positive and the addition of the negative-going sync pulse causes the total voltage to swing to the negative polarity, allowing the transistor to conduct and switch on. Any definite arrangement of the sync pulse will be sufficient to maintain reasonable stability in the astable multivibrator. In other words, it is not necessary to pulse each cycle of the square wave or even every other cycle; every fifth, tenth, or other regular triggering interval should be ample.

Designing a Collector-Coupled Astable Multivibrator

Multivibrators used as clocks in digital computers must operate at high frequencies; therefore, design problems do become quite complex. Most medium or lower frequency circuits, however, are rather simple. To design a free-running multi to produce a square wave of 10 kHz first requires the selection of the supply voltage (Vcc). This value is not critical, but normally would be somewhat less than half the maximum voltage rating of the transistor being used. Using a pair of PNP computer-type transistors (2N1305) with a maximum voltage rating of 35v, we could choose a 9v transistor-type battery for Vcc. Collector resistors R1 and R2 (Fig. 2-3) should be equal in value; two other primary considerations are the impedance of the load to be handled and collector current Ic.

The lower the value of the collector resistors, the greater the load we can accommodate without affecting the rise time of our square-wave output. However, lower values cause more collector current to flow, which reduces battery life (Vcc) since, when each transistor is on, current through it is limited only by the resistor. Ic equals Vcc divided by R1, and this formula applies to both transistors. When selecting component values a compromise would seem to be the most reasonable solution—low enough to avoid distortion of rise time and high enough to bring battery drain within practical limits. Choosing a collector resistor value of 2700 ohms provides an Ic of little more than 3 ma, yet permits a low-load impedance without adverse effects. Biasing resistors R3 and R4 should not exceed beta X R1 to insure that the on or conducting transistor is turned on completely. Values of 10K to 50K are normally applicable in multivibrators, but the limits are dictated either by the sensitivity to temperature variations present with high resistance values or by using larger, more costly capacitors with low values. Since the beta for the 2N1305 transistor is 40 or more (actually to 200), beta X R1 equals 40 X 2.7 or 108K. A compromise of about one third on the bias resistor values gives us a standard of 33K.

The coupling capacitor values are selected by the formula, C equals 1 divided by 1.4Rf; R is the value of R3 or R4 and f the frequency of oscillation chosen. Since our oscillation frequency is to be 10 kHz, the capacitors for our square-wave generator would be:

$$C \text{ equals } \frac{1}{1.4 \text{ X } 33 \text{ X } 10^7} \text{ equals .002164 mfd}$$

The closest standard value of capacity is .0022 mfd.

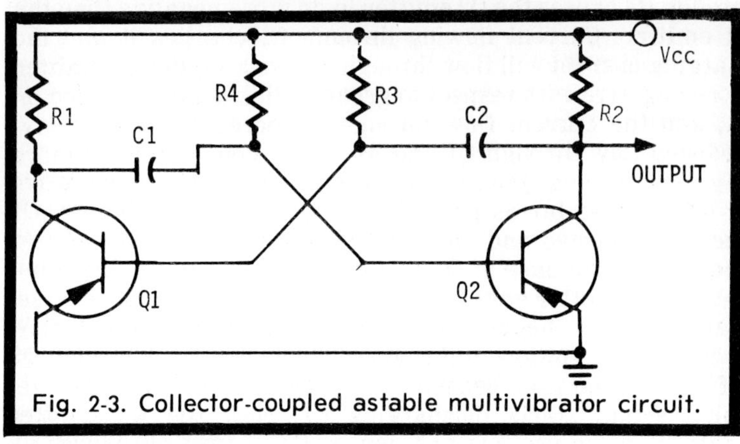

Fig. 2-3. Collector-coupled astable multivibrator circuit.

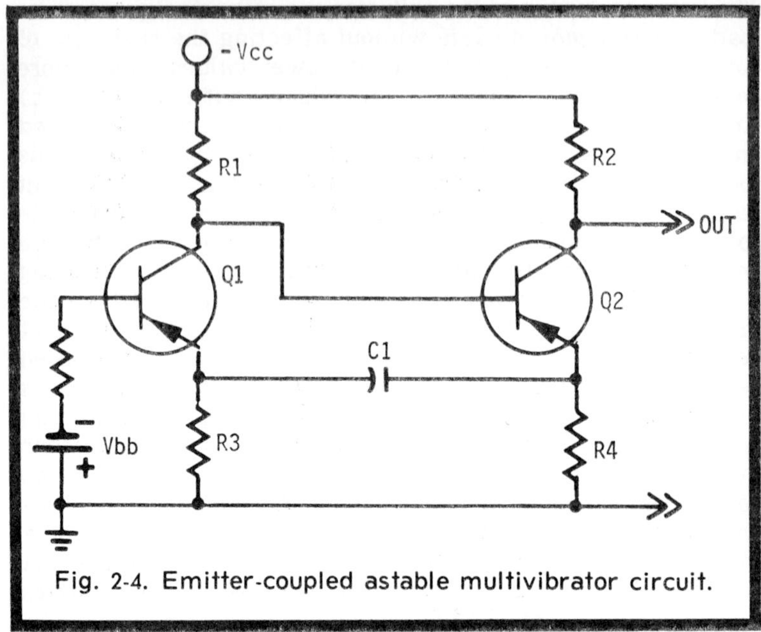

Fig. 2-4. Emitter-coupled astable multivibrator circuit.

Emitter-Coupled Astable Multivibrator

Only one coupling capacitor is required in the emitter-coupled astable multi, and this fact makes frequency changes a much simpler matter. However, it is definitely more difficult to adjust for normal operation. In Fig. 2-4, R2 is lower in value than R1 in the emitter-coupled circuit, thus Q2 is forward-biased through R1 and will conduct more heavily. Current flow through R4 causes the Q2 emitter to go more negative than the Q1 emitter. Current flowing through C1 charges it and the charging current will flow through R3, making the Q1 emitter more negative with respect to ground. Bias is now reduced on Q1, and the current flow through it decreases accordingly. Reduced flow through Q1 causes the Q1 collector to go more negative, turning Q2 on harder. The action is now regenerative as Q2 is on as far as possible and Q1 is off. Continuing, C1 becomes charged and the negative voltage at the Q1 emitter decreases. The base of Q1 is connected to bias battery Vbb, and as the emitter voltage approaches -Vbb, Q1 is not reverse-biased and so begins to conduct. Increased current flow decreases the voltage at the Q1 collector and starts turning Q2 off. The Q2 emitter becomes less negative and the positive-going signal starts discharging C1. Now the Q1 emitter becomes more positive and causes Q1 to turn on harder as

28

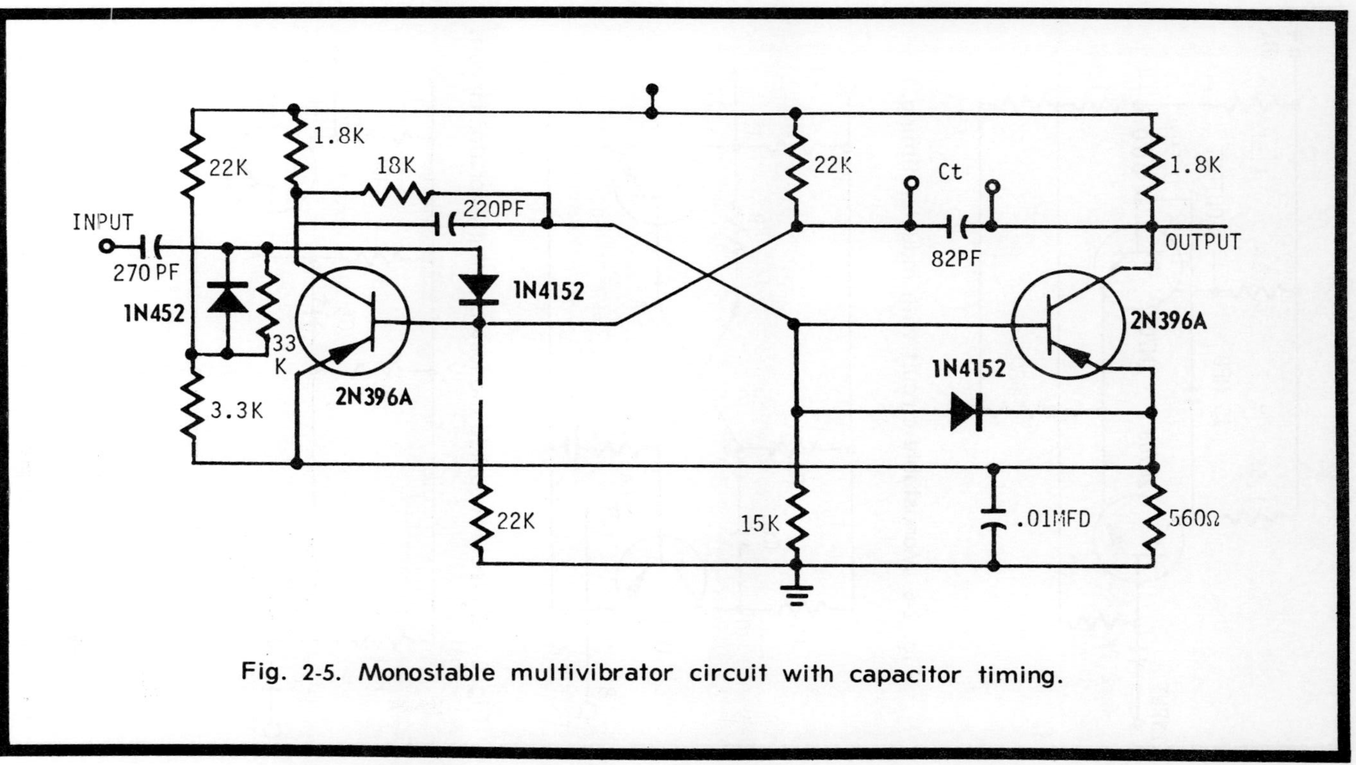

Fig. 2-5. Monostable multivibrator circuit with capacitor timing.

29

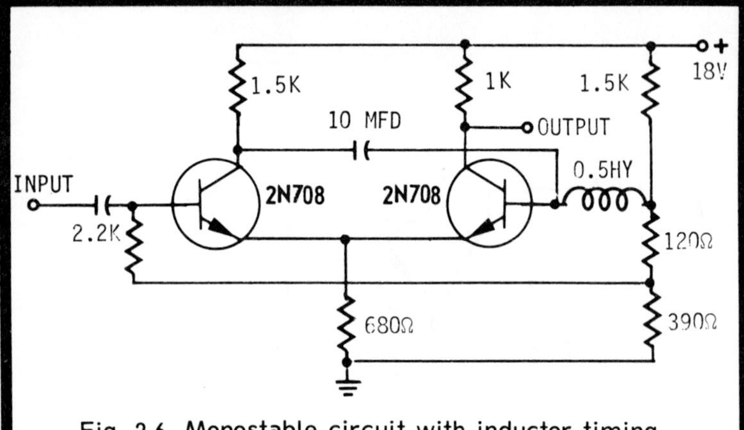

Fig. 2-6. Monostable circuit with inductor timing.

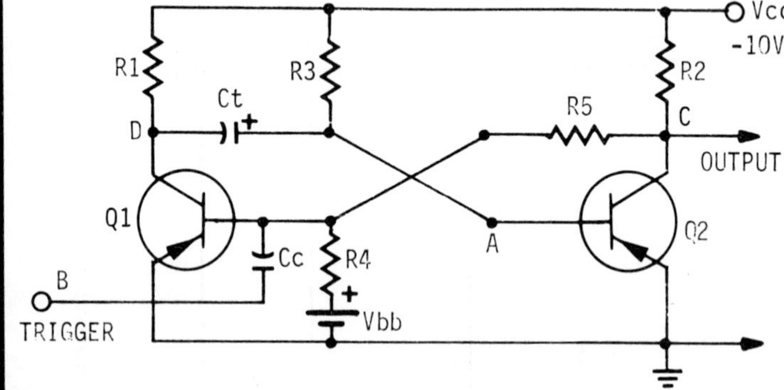

Fig. 2-7. Collector-coupled monostable multivibrator circuit.

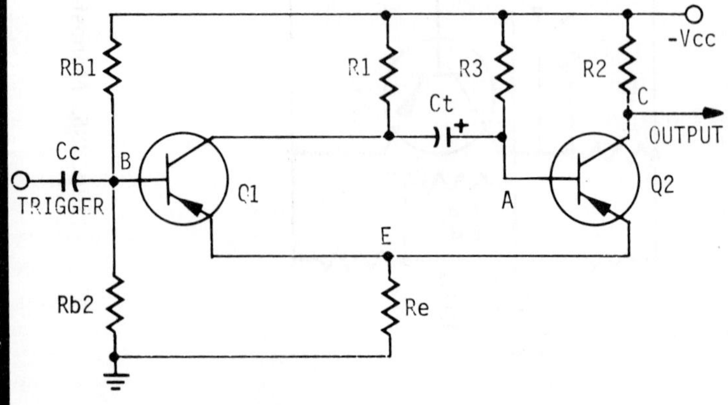

Fig. 2-8. Emitter-coupled monostable multivibrator circuit.

regeneration takes place, cutting off Q2 as Q1 reaches saturation. C1 now discharges through Q1 and R4 as the Q2 emitter goes more positive than the Q2 base, and Q2 begins to conduct as the cycle repeats.

Monostable Multivibrator

By simple triggering action the monostable multivibrator switches to the unstable state and there remains for a pre-selected time before returning to the original stable condition. Thus the monostable circuit is suitable for standardizing pulses of random widths or for generating time-delayed pulses. The circuit parallels the flip-flop or bistable multi except that one cross-coupling network permits AC only to be coupled. Hence, the flip-flop can remain in its unstable state only until the circuit reactive components discharge. Fig. 2-5 illustrates a monostable circuit with capacitor timing and Fig. 2-6 shows the use of inductor timing. The inductor, as might be expected, gives much better pulse-width stability at high temperatures.

Collector-Coupled Monostable Multi

As the diagram of Fig. 2-7 shows, Q2 is forward-biased through resistor R3. Its collector is near ground potential since it is fully on. Q1 is reverse-biased by Vbb and the stable condition of the circuit has Q1 off and Q2 on. Ct becomes charged through the forward-biased base-emitter diode of Q2 and the circuit remains in this state until triggered. As a negative trigger pulse reaches the base of Q1, it is turned on and the collector begins to go less negative with a positive-going pulse being fed to the base of Q2 as a result of the charge on Ct. Q2 now begins to turn off as a negative-going signal is fed to the base of Q1, turning it on harder. Regeneration begins and Q1 rapidly saturates as Q2 is cut off. Q2 remains cut off as a result of the charge on Ct, and as the latter discharges through R3 the base of Q2 begins to go negative as R3 is returned to -Vcc, forward-biasing Q2 and starting to turn it on. Regeneration causes Q2 to switch on completely and Q1 is now off.

Since Q2 is in saturation and Q1 is reverse-biased by +Vbb, there is nothing to cause Q1 to turn on again and the circuit remains thus, awaiting the next trigger pulse. Taking the output at the Q2 collector, the rectangular output pulse has a time duration of approximately 0.7R3Ct. Since capacitor Ct has charged through the forward-biased base-emitter junction of Q2, the charge is about equal to supply voltage Vcc or 10v. When the trigger is applied, it switches Q1 on and the latter goes into saturation, appearing as a short circuit to ground at its collector. With Ct charged to 10v, the base of Q2 is positive instantly as Q1 turns on. Now Q2 is reverse-biased and cut off with its base appearing as a very high resistance (nearly open), so Ct begins to discharge through R3. R3 is connected to -10v and the voltage at the Q2 base begins to drop from +10v to -10v. Ct would require five time constants to change from +10v to -10v without Q2 in the circuit, but as it is so situated it switches on immediately as the voltage on the base of Q2 begins to go negative. Q2 is on as the voltage at its base crosses zero on the way from +10 to -10. The universal time constant chart shows that the half-way point between the capacitor's initial and final voltage values is reached at 0.7 time constant, placing the duration of the pulse at 0.7RC.

In the astable multivibrator, the period of one cycle of the square wave was 1.4RC, because the astable has two quasi-stable states each having a duration of 0.7RC. The trigger is required only to start the astable; after that the trigger pulse is no longer needed. This means that the trigger may be very short in duration compared to the output pulse. As proper operation depends on the signal fed from the collector of each transistor to the base of the other, it is called collector-coupled.

Emitter-Coupled Monostable Multivibrator

Another form of one-shot multi coupling simply provides a different connection to the circuit, resulting in one stable state and one quasi-stable state, just as offered by the collector-coupled version. In the stable condition voltage at the common emitters is negative as a result of current flow through Re and Q2, forward-biased by R3. Rb1 and Rb2, as shown in Fig. 2-8, produce a voltage difference between the Q1 base and the Q1-Q2 emitter to reverse bias Q1 and, therefore, cut it off. When the trigger pulse is applied to the base of Q1, it begins to turn on, and as the collector goes less negative a positive signal reaches the base of Q2 due to the charge on Ct. This causes Q2 to turn off, and because Q2 has a smaller collector resistor than Q1, current flow through common-emitter resistor Re will be less when Q1 is conducting than if Q2 is conducting. Therefore, voltage at the emitter is less negative with Q2 cut off and Q1 on. As Ct discharges through R3 with Q2 off, the voltage at the emitter is less negative than at the Q1 base, causing Q1 to remain on. The charge of Ct continues to decrease until Q2 starts to conduct again, and as this takes place the emitter connection becomes more negative and Q1 is cut off. This being the stable state of the circuit, it is held until the next triggering pulse comes along.

Monostable Multivibrator Uses

Rectangular pulses of pre-determined duration, which begin with a triggering pulse, offer many useful applications where a pulse generator with good synchronization is required. The monostable multi is frequently used to unblank the CRT of a scope during sweep time, as the sync pulse that starts the sweep also triggers the one-shot. This causes an output pulse to be applied to the grid of the CRT, reducing the cut off bias and allowing the electron beam to reappear on the screen during horizontal sweep. As the pulse of the monostable is accurately timed to coincide with the sweep time, the beam is cut off at the end of the horizontal sweep during retrace and as the next triggering pulse starts the sweep again, it also triggers the multi to again unblank the CRT and thus permit the trace to appear on the screen.

The monostable circuit finds its most important use in time-delay functions and, compared to delay lines, offers the features of smaller size, lower cost, and easier pulse-duration adjustment. In delaying a trigger pulse by any preset number of microseconds, say, 20 us, the positive triggering spike is applied to the one-shot multi and a 20 us rectangular pulse appears at the output of the previously adjusted unit. By feeding this output into a differentiator, we have a negative spike followed by a positive spike 20 us later. As only the single positive spike is desired, this differentiator output is passed through a clipper which delivers the single positive spike at precisely 20 us after the original trigger was applied to the monostable input. The spike thus reaches the load at $t+20$ us.

Gating circuit applications in the computer field offer additional functions for the monostable device when the passing of a pulse within a specific period of time following a timing pulse is a fundamental requirement. Applying the trigger pulse to the monostable multi (gating circuit) adjusted for the pre-selected pulse duration of, say, 20 us, the output pulse of that duration is passed to the AND gate. This has the effect of keeping the gate open for a period of 20 us by assuring passage of any pulse of like polarity that may reach the AND gate during that period. Any pulse at the AND gate during any other time or of opposite polarity would not be seen at the output of the AND gate.

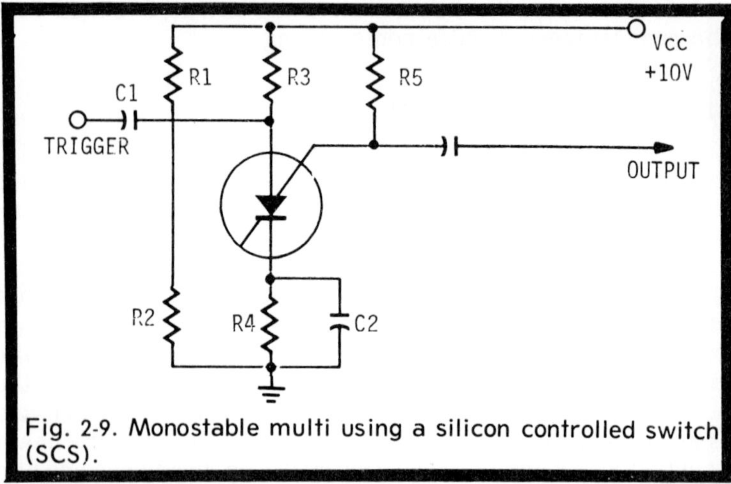

Fig. 2-9. Monostable multi using a silicon controlled switch (SCS).

SCS Monostable Multivibrator

A unique feature of the SCS is its sensitive triggering capability, an advantage that may be utilized in many ways. At moderate temperatures where leakage current is not excessive, the use of high triggering source impedances without objectionable loading is possible. The anode gate offers a versatility that is lacking in the SCR, and the rapid turn-on to maximum conduction is useful. Control of positive feedback is made possible by the anode gate, and the potential gain of the device exceeds 100 within acceptable limits of stability. The use of the silicon controlled switch in the one-shot configuration takes advantage of the low idling current as the device is off during this period. By applying a positive pulse to the anode coupling capacitor C1 (Fig. 2-9), the SCS is turned on. As soon as capacitor C2 charges sufficiently, the SCS is returned to its original state and capacitor C2 discharges through R4. The multi remains off until the next triggering pulse, at which time the cycle is repeated. Planar devices handle higher voltages and currents at lower cost and are preferred for late design. The lower internal resistances in these units offer a distinct advantage in gate turn-off and dissipation characteristics. The monostable multi in Fig 2-9 uses a planar type of the silicon controlled switch.

Bistable Multivibrator

The bistable multivibrator is actually a variation of the astable except that the base-biasing resistors are returned to a

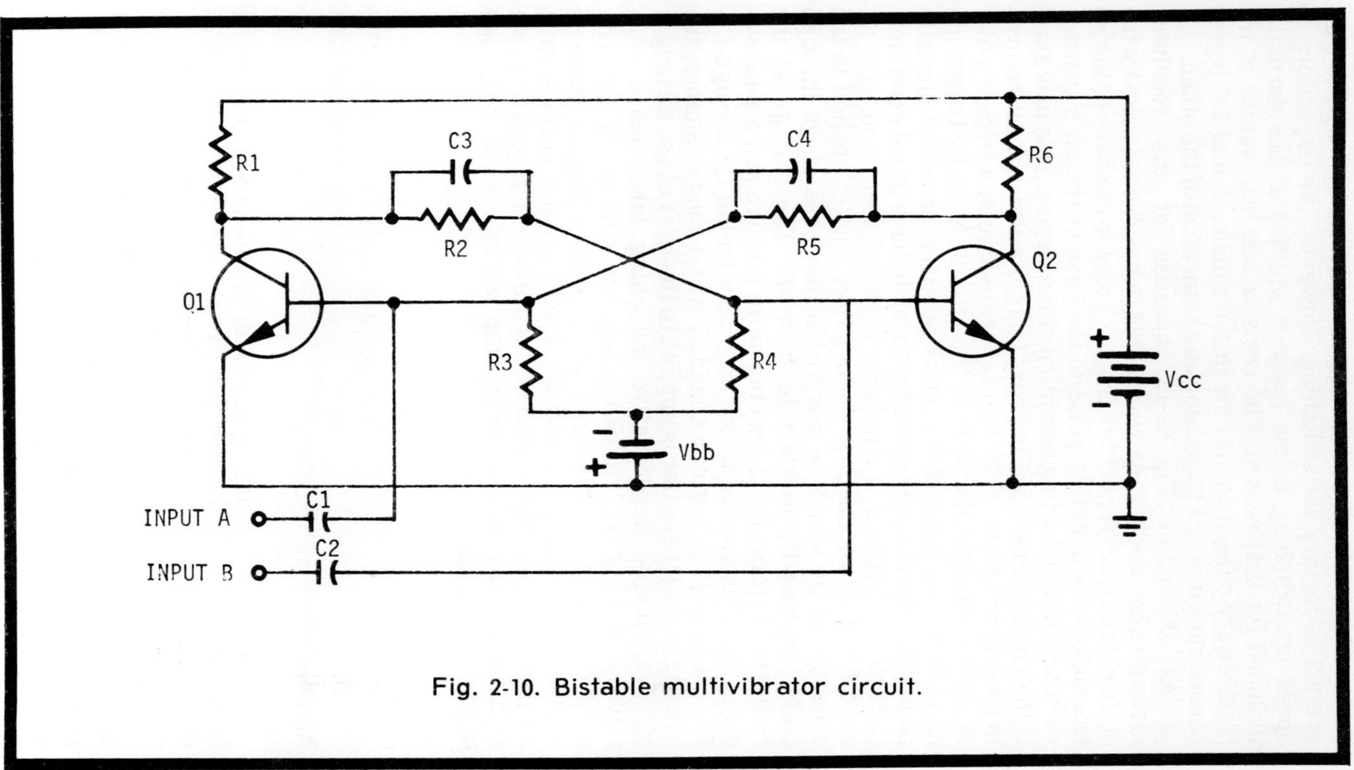

Fig. 2-10. Bistable multivibrator circuit.

reverse-bias voltage and the capacitors from collector to base are not necessary for switching. However, the capacitors do speed up switching in the bistable multi and are normally included for that reason. The resistor and bias values of the bistable are chosen so that the initial application of DC power causes cut-off of one transistor and saturation of the other.

As shown in Fig. 2-10, because of the feedback arrangement, each transistor is held in its original state by the state of the other. A positive trigger pulse to the base of the off transistor or a negative pulse to the base of the on transistor switches the conducting condition of the circuit. This new state is then maintained until the next trigger pulse when it returns to its original state. With two separate inputs, a trigger pulse at input A changes the state of the circuit. Now, by triggering input B with a pulse of the same polarity, or one of opposite polarity at input A, the circuit returns to its original condition. Collector triggering is similarly accomplished, and with Q1 cut off and Q2 conducting, the negative trigger applied to the collector of Q1 would be coupled to the base of Q2 through C3R2, causing Q2 to start to turn off. The voltage at the collector of Q2 begins to become more positive and a positive-going signal is applied to the base terminal of Q1 through C4-R5, causing Q1 to begin to conduct. As Q1 conducts more, the voltage at the Q1 emitter is applied to the base of Q2, turning it off even more. Regenerative switching takes place very rapidly, driving Q1 into saturation and cutting off Q2. The circuit remains in this state until another trigger pulse is applied. Capacitors C3 and C4 speed up the regenerative switching action. The output of the circuit is a square wave with continuous pulsing or a step voltage when only one trigger is applied.

BLOCKING OSCILLATOR

Astable Blocking Oscillator

In the astable blocking oscillator circuit in Fig. 2-11, Q1 is forward-biased. By closing the switch, the transistor is forward-biased through -Vcc and the primary of T1. Although the collector current is zero momentarily, it must rise very quickly when the circuit to Vcc is completed. This causes a heavy current flow in the secondary which in turn increases Ib and drives Q1 wide open, simulating a closed switch. As Q1 is such a low resistance, the voltage at C is about zero and that across the primary equals Vcc.

36

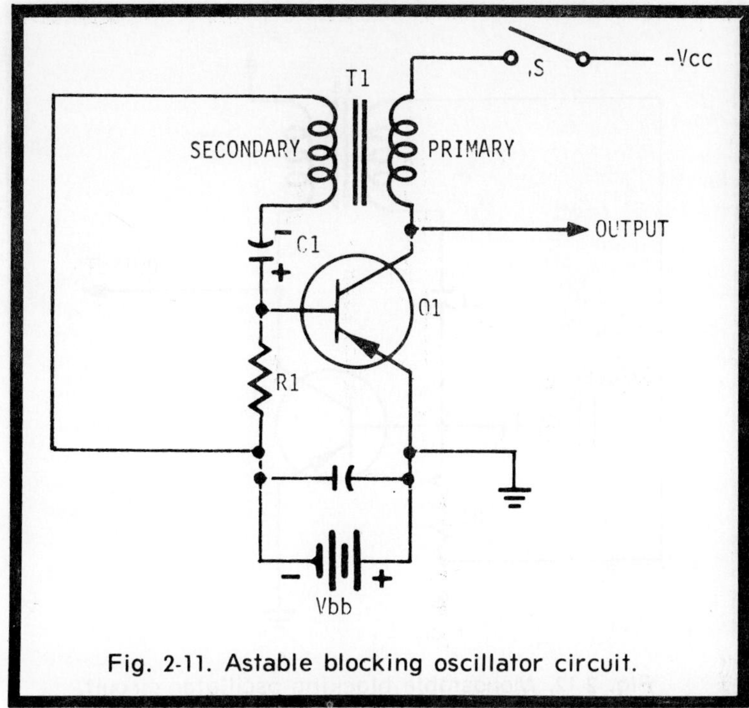

Fig. 2-11. Astable blocking oscillator circuit.

Base current, Ib, remains constant only while the collector current rises at a constant rate, causing the induced current to remain steady. Although the collector current is equivalent to Ib X beta, it requires time to reach this value as the rise must not be too fast. The induced voltage rises with the current and cannot be greater than Vcc. Since the collector does reach a current value of Ib X beta, it stops rising because it can never exceed this value. Q1 ceases to be saturated and no longer acts as a switch but rather as an amplifier.

When Ib X beta halts the collector current rise, the voltage induced in the secondary falls to zero and drops the base current which results in a fast drop in collector current. With the decrease in collector current, the induced voltage is opposite in polarity to that induced with the rising collector current. This causes the base of Q1 to be reverse-biased and cuts off the collector current. Now the induced voltage in the T1 secondary collapses completely.

Capacitor C1 becomes charged during the time of base current flow; when it discharges a voltage drop appears across R1, reverse-biasing the base until C1 discharges enough for Vbb to again take control and forward-bias Q1 to begin another cycle.

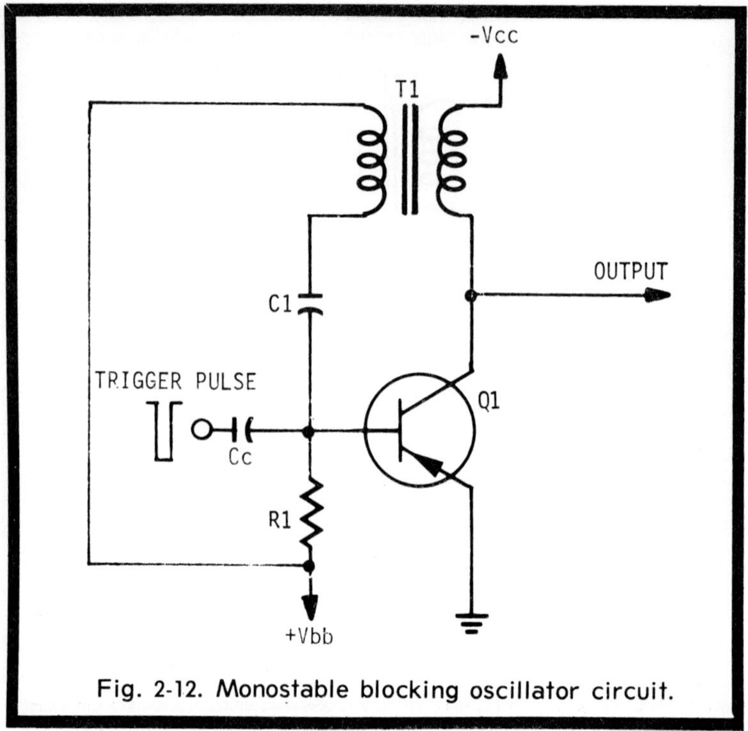

Fig. 2-12. Monostable blocking oscillator circuit.

Monostable Blocking Oscillator

As is the case with the monostable multivibrator, the monostable blocking oscillator presents a single output pulse for each triggering pulse. Fig. 2-12 shows a one-shot blocking oscillator circuit identical to the astable type except that Q1 is off due to reverse-bias.

Applying a triggering pulse to the base of Q1 through capacitor Cc causes the transistor to start conducting. Current flowing in the primary of T1 induces a voltage in the secondary of that transformer, and positive feedback through C1 to the base of Q1 drives the transistor into saturation. The rest of the cycle parallels the astable blocking oscillator but for the fact that C1 discharges only to the positive voltage level established by Vbb. Q1 remains at cut-off until the next trigger.

Blocking Oscillator Output

The output may be taken from the transistor collector terminal or, if a signal of opposite polarity is desired, from the high side of a resistor in the emitter lead. If isolation is

38

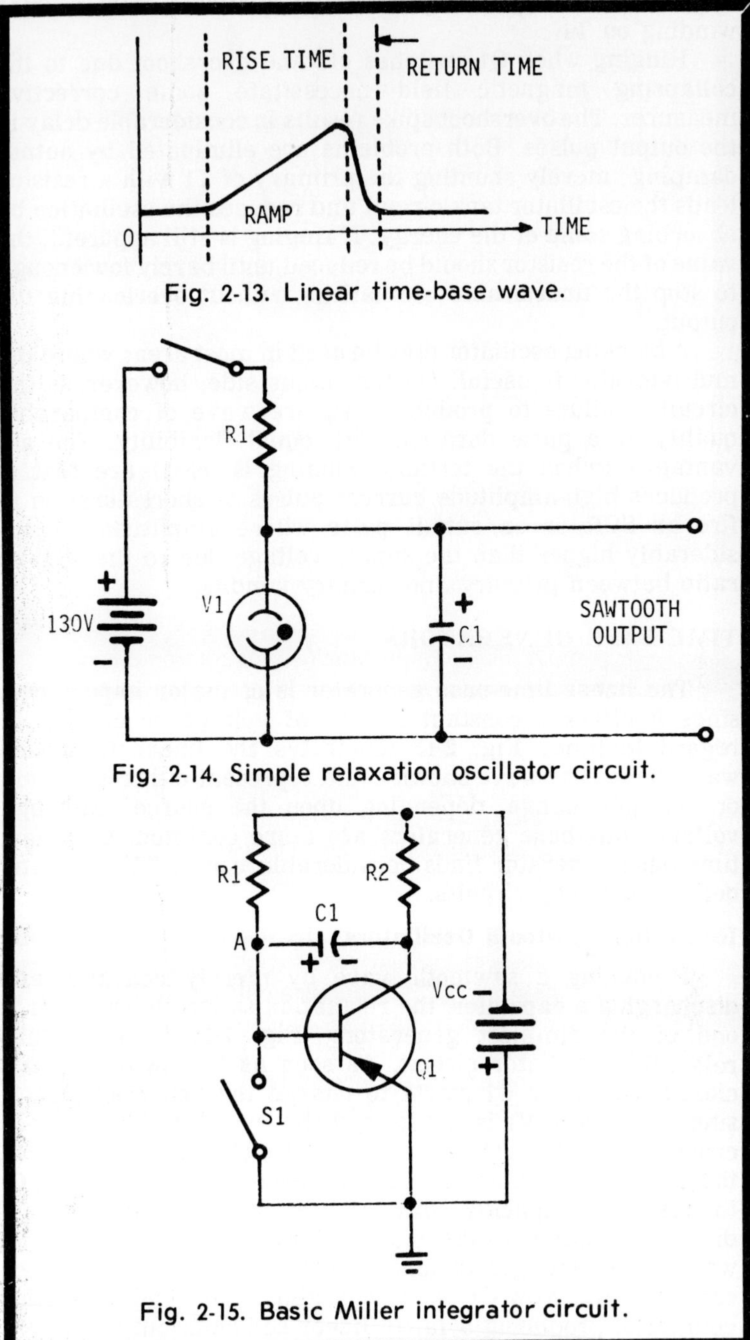

Fig. 2-13. Linear time-base wave.

Fig. 2-14. Simple relaxation oscillator circuit.

Fig. 2-15. Basic Miller integrator circuit.

required, the output may be made available from a third winding on T1.

Ringing when Q1 switches off and overshoot due to the collapsing magnetic field necessitate some corrective measures. The overshoot spike results in considerable delay in the output pulses. Both problems are eliminated by output damping; merely shunting the primary of T1 with a resistor loads the oscillator tank circuit and reduces the oscillation by absorbing some of the energy. If ringing is still apparent, the value of the resistor should be reduced until barely low enough to stop the undesirable oscillations without overloading the output.

A blocking oscillator may be used in most areas where the multivibrator is useful. On the minus side, however, is the circuit's failure to produce a square wave of comparable quality or a pulse duration with equal flexibility. The advantages (when the tertiary winding is used) are that it produces high-amplitude current pulses of short duration to fire an SCR or an output pulse whose amplitude is considerably higher than the supply voltage due to the step-up ratio between primary and tertiary windings.

TIME-BASE GENERATORS

The linear time-base generator is of major importance, since it offers a constant change of voltage or current in regard to time; Fig. 2-13 illustrates the linear time-base waveshape. The wave shown could represent either a current or voltage change, depending upon the source. Although voltage time-base generators are more common, a current time-base generator finds considerable use in TV magnetic deflection sweep circuits.

Relaxation Sawtooth Oscillators

Producing a sawtooth wave by merely charging and discharging a capacitor, the relaxation sawtooth oscillator is one of the simplest generators. Fig. 2-14 is a gas-tube relaxation oscillator circuit. As soon as the switch (S1) is closed, capacitor C1 starts to charge through resistor R1; since the gas in V1 is not ionized the tube acts like an open circuit. When the voltage across V1 reaches the firing level, the gas ionizes and presents a very low resistance, causing C1 to discharge quickly through this path. However, C1 discharges only until the extinguishing voltage is reached, at which point the gas de-ionizes and appears to be an open circuit again. Now C1 starts charging again and oscillation continues, producing a fairly linear sawtooth output.

40

Miller Integrator

In order to produce the better linearity required in many applications, the Miller sweep circuit, in any of its many variations, produces a time base with improved qualities by discharging a capacitor through a transistor and using feedback to the base to maintain the discharge rate at a constant level. The constant-current charging or discharging of a capacitor assures that the voltage across it will increase or decrease linearly with time. Since the rate of voltage change is constant, the ramp must be linear. V equals Q divided by C and Q equals I_T, so by maintaining current I constant with the increase in time, charge Q on the capacitor must increase directly with that time. The voltage across the capacitor equals the charge divided by the capacitor value. The voltage increases directly with time.

Referring to the simplified Miller integrator circuit in Fig. 2-15, during the time period t1 to t2, C1 discharges to produce the sawtooth ramp; after t2, capacitor C1 charges again. When switch S1 is open, the capacitor discharges, forming the ramp. When S1 is closed the ramp drops off as C1 charges once more. The Miller circuit must be externally triggered, depending on pulses to open and close S1. Transistor Q1 is cut off when S1 is closed and capacitor C1 charges through R2 at this time.

Feedback and Equilibrium

When a transistor is not in saturation, collector current Ic must equal Ib times beta. Adjusting Ib permits any desired value of Ic, regardless of the value of the collector voltage. The adjustment of Ib is a direct function of the feedback from the collector through C1 to the base, thus controlling the value of base current Ib for the necessary Ic value.

If VC1 is 5v, the output voltage at Point A in Fig. 2-15 must be -5.2v. So if Ib did not increase as the capacitor discharged, Ic could not increase either, and neither could IR2. With the voltage drop across R2 frozen, the output voltage at A is likewise pegged at -5.2v. When capacitor C1 discharges to 4.9v, the base voltage shifts from -0.2v to -0.3v and such a change at the base causes Ib to increase considerably. The slightest shift in output voltage from normal for that time of the discharge period brings about an equal change in base voltage, causing a change in base current and thus in collector current, until Ic is correct for maintaining the output voltage exactly 0.2v more negative than VC1.

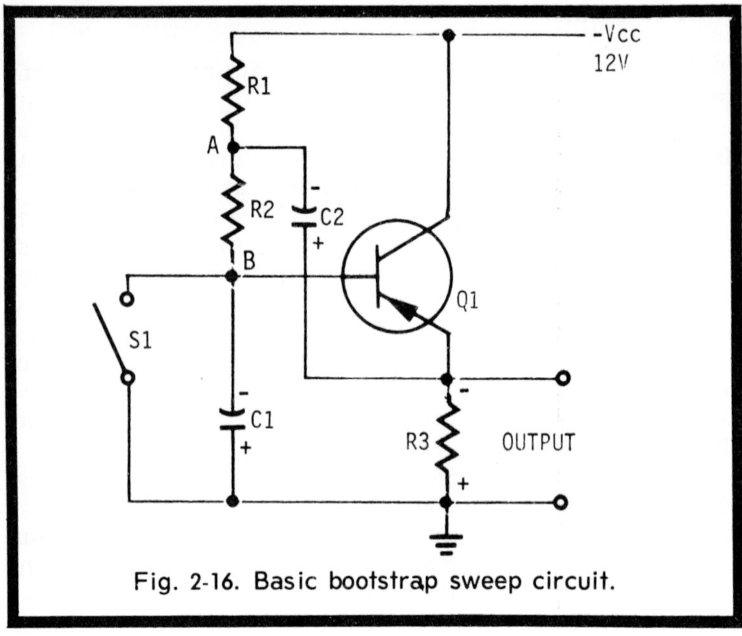

Fig. 2-16. Basic bootstrap sweep circuit.

The capacitor discharge current Id is constant—resistor R1 divides between Ib and Id—and with Ib constant, Id must follow. However, Ib increases as C1 discharges, but with high-gain transistors this increase is limited and the percent change in discharge current Id is negligible due to the gain. This negligible change in discharge current may be further reduced by increasing transistor amplification to provide even more gain.

Bootstrap Time-Base Generator

Using feedback in the emitter-follower configuration, the linearity of the ramp is greatly improved as shown in Fig. 2-16. With the switch open, capacitor C1 charges via R1 and R2, forming the ramp of the sawtooth as the voltage increases across C1. The feedback circuit from Point A through C2 to the emitter improves the linearity of the bootstrap. Like the Miller circuit, the bootstrap sweep circuit requires an external gating signal to begin the sweep. In order to stop the sweep and return the output to its proper point before the capacitor is fully charged or discharged, the gating circuit is required. For this gating circuit, the one-shot multivibrator is quite satisfactory. Since NPN transistors are needed to provide the positive-going output pulse, and a 10 ms pulse duration is required, the RC timing circuit values must be carefully selected.

Current Time-Base Generator

A current time-base generator, frequently used in TV receiver deflection circuits generates a current through the yoke that varies in a predetermined manner with time. The linearly increasing current flows as long as there is little resistance in the coil or inductance. The linearly increasing magnetic field sweeps the electron beam across the CRT screen at a constant rate and thus avoids stretching or shrinking of portions of the picture as is observed when linearity problems exist in these circuits.

The Darlington circuit shown in Fig. 2-17 eliminates difficulties frequently encountered in less sophisticated circuitry. In order to avoid problems involved in simple switching, special circuits are used to force the current to flow through the yoke at a predetermined rate. By virtue of its high impedance, which does not distort the input ramp, the Darlington circuit is merely two cascaded emitter-followers

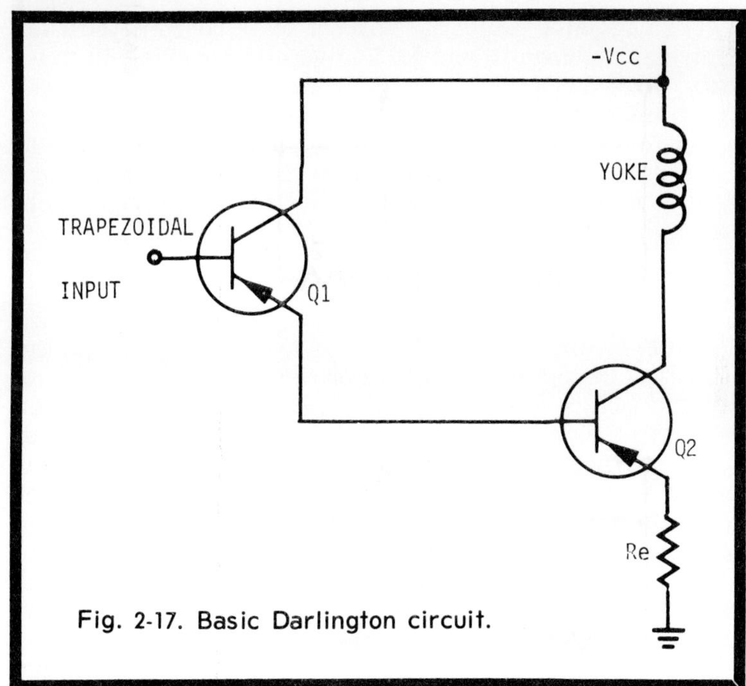

Fig. 2-17. Basic Darlington circuit.

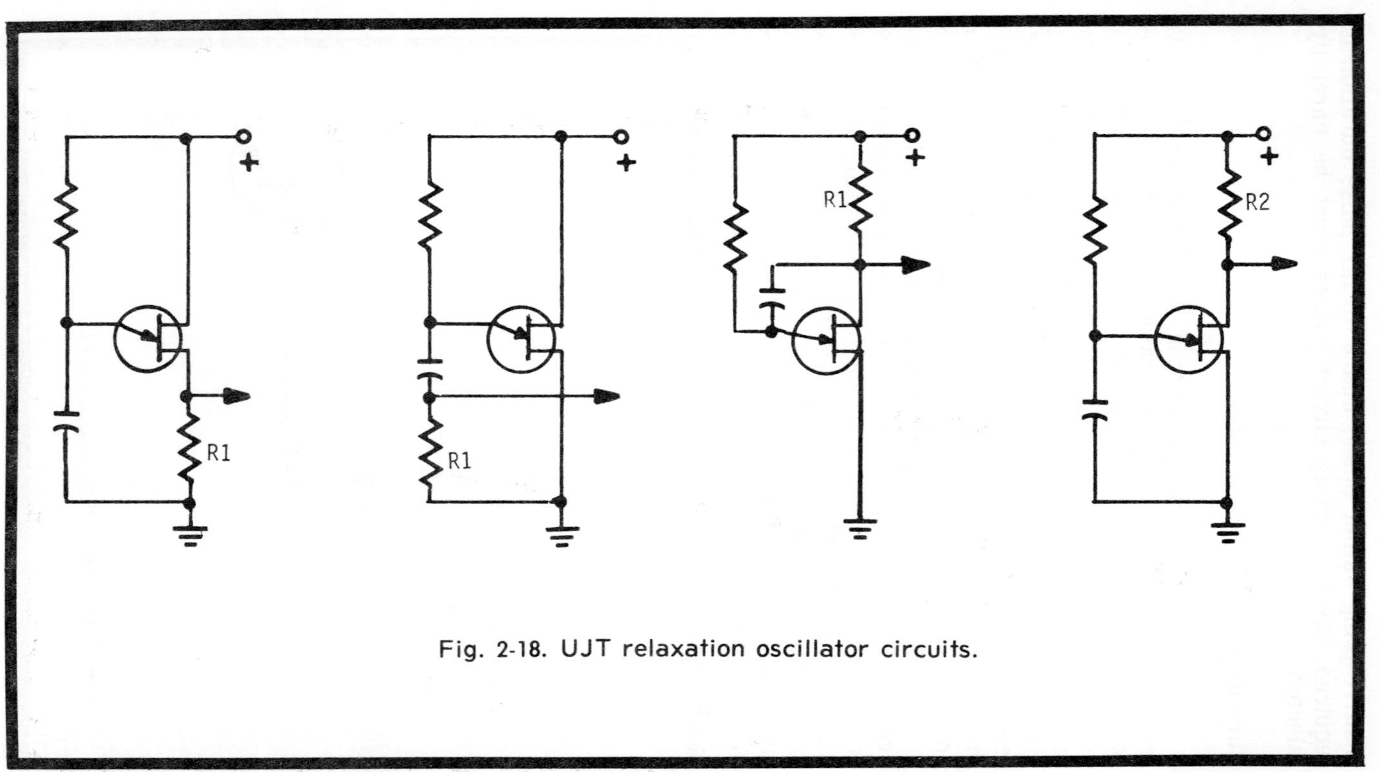

Fig. 2-18. UJT relaxation oscillator circuits.

having an extremely high input impedance equivalent to the multiple of the betas of Q1 and Q2 times Re. The input pulse is trapezoidal, consisting of a step and a ramp. The step does reduce the effect of the coil resistance and the current through Q2 is controlled as desired by the input wave. The current through Q2 controls the current through the yoke, and the rate of change of the input voltage ramp directly regulates the rate of change of the current through the deflection coil.

UJT PULSE GENERATORS

Relaxation Oscillator

When a UJT relaxation oscillator conducts, a current pulse flows in the emitter, base-one, and base-two circuits. The relaxation oscillator is an efficient pulse generator, capable of generating either negative- or positive-going pulses at suitable impedance levels. Fig. 2-18 shows various

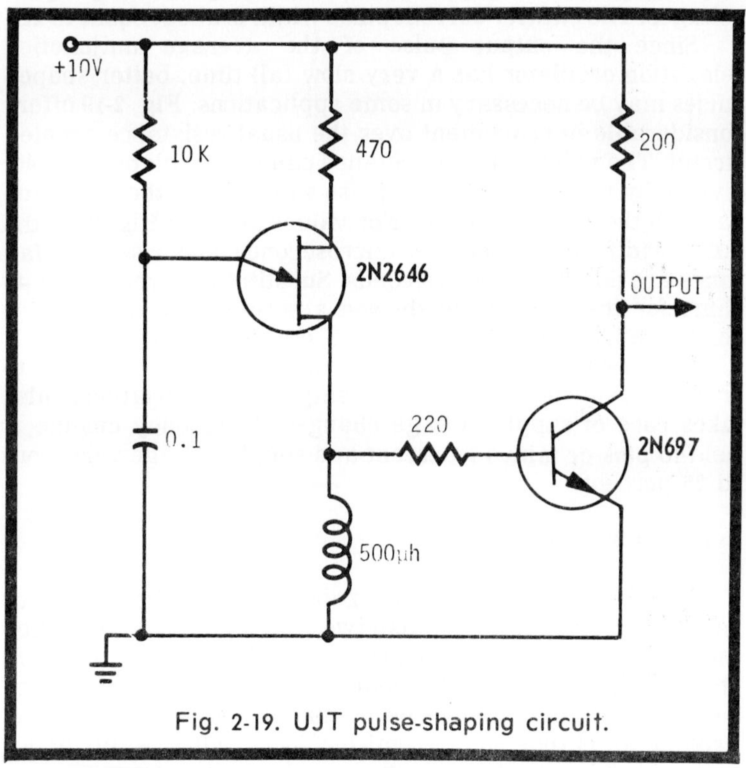

Fig. 2-19. UJT pulse-shaping circuit.

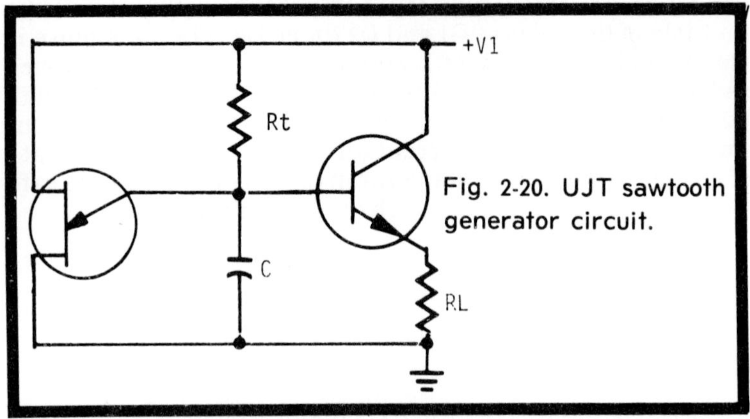

Fig. 2-20. UJT sawtooth generator circuit.

unijunction pulse generator circuits. Since three of the circuits use the discharge current of the capacitor to form the pulse, the output impedance is low, and the final circuit, using base-two current to provide the pulse, presents a much higher output impedance and has the capability of generating higher voltages. The emitter current pulse width is about equal to 2tf.

Since the output pulse of the average unijunction relaxation oscillator has a very slow fall time, better shaped pulses may be necessary in some applications. Fig. 2-19 offers considerable improvement over the usual resistance-coupled circuit. The value of the inductance can be calculated by $0.4t^2$ divided by C; t is the desired pulse width and C the capacitor value in the emitter circuit. For values given in Fig. 2-19, the pulse width approaches 12 microseconds and rise and fall times are about 0.3 microseconds. Substituting a resistor of 47 ohms for the inductance showed a similar rise time, but a considerably slower fall time of 3 microseconds.

Base-two resistor R2, although normally chosen to prevent frequency shifts with temperature variations, also takes care of supply voltage changes. Frequency change is held to plus or minus 1 percent and supply voltage variations to 25 percent.

Synchronization

By means of either positive-going pulses at the emitter or negative-going pulses at base-two, a unijunction relaxation oscillator is readily synchronized. The pulses for synchronization must be high enough in amplitude to reduce the peak point voltage below the instantaneous emitter voltage. Trigger amplitude at the emitter for a pulse width of 0.5 microseconds would be +1.7v or at base-two -2.3v.

46

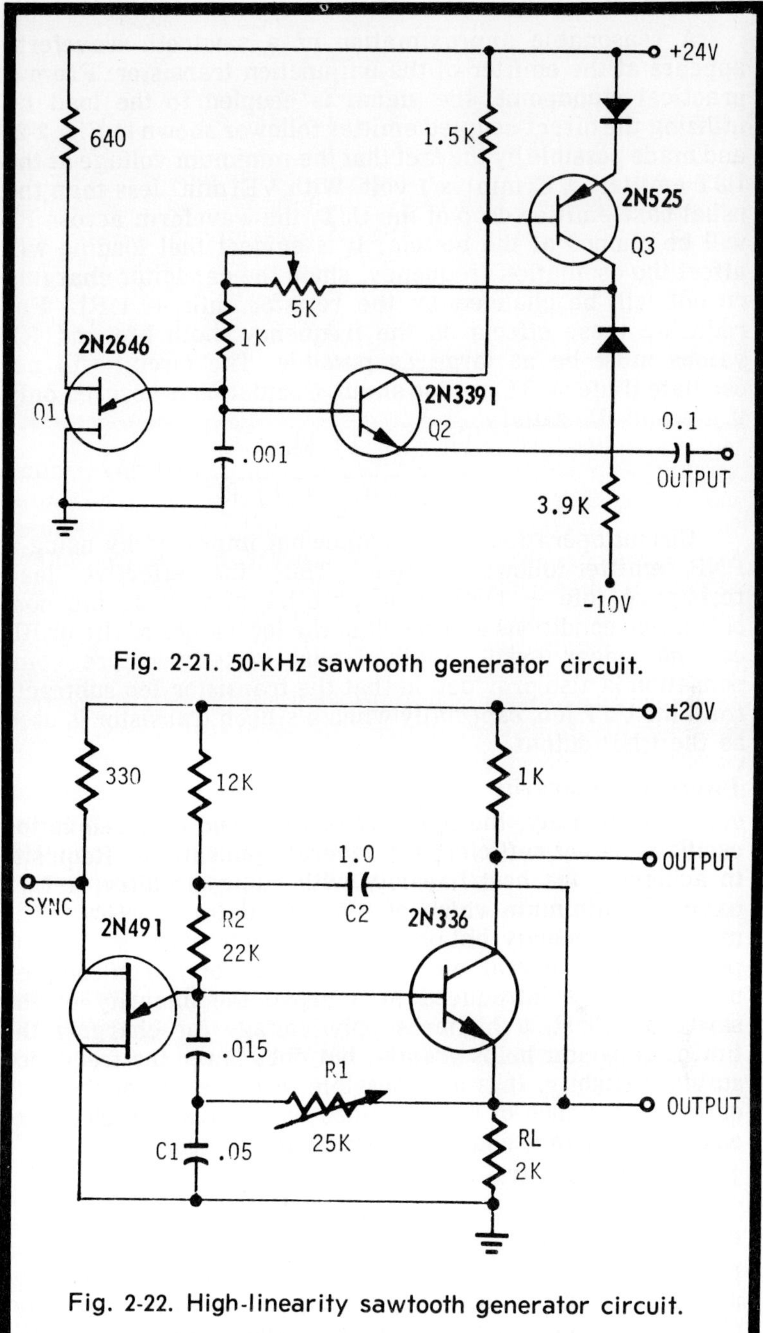

Fig. 2-21. 50-kHz sawtooth generator circuit.

Fig. 2-22. High-linearity sawtooth generator circuit.

Sawtooth Generators

A reasonable approximation of a sawtooth waveform appears at the emitter of the unijunction transistor. From a practical standpoint, the signal is coupled to the load by utilizing the direct-coupled emitter follower shown in Fig. 2-20 and made possible by the fact that the minimum voltage at the UJT emitter, VE(min) is 1 volt. With VE(min) less than the usual base-emitter drop of the UJT, the waveform across R1 will be clipped at the bottom. It is evident that loading will affect the oscillation frequency, since the capacitor-charging circuit will be changed by the resistor (hfe + 1)RL. For reducing these effects on the frequency, both hfe and RL values must be as large as possible. The circuit will not oscillate if hfe or RL are too small. Oscillation is ensured only if hfe and RL satisfy:

$$\frac{(h_{fe} + 1)\ RL}{Rt + (h_{fe} + 1)\ RL}\ n(MAX)$$

Circuit operation can be somewhat improved by using a PNP emitter-follower output. Thus the effective load resistance (hfe + 1)RL is in parallel with Rt so the non-oscillation conditions as a result of the low values of Hfe or RL can no longer exist. Another plus in temperature compensation is also provided in that the transistor Ico subtracts from the UJT Ieo, especially when a silicon transistor is used as the PNP output.

Improving Linearity

Unfortunately, the linearity of the basic UJT relaxation oscillator is not sufficient for several applications. Requisite to achieving the best linearity with basic circuitry, a UJT having a minimum value of n is mandatory (2N489). The maximum linearity obtainable with such types is near 10 percent, and several circuit techniques of a non-complex nature can be introduced to improve the linearity of the sawtooth. First, a higher supply voltage for charging the timing capacitor helps greatly, but does lower the frequency stability slightly. It is also possible to make use of the high output impedance of the grounded-base transistor circuit to ensure a constant charging current for the capacitor.

Linear Sawtooth-Wave Generators

A bootstrap charging circuit is illustrated in Fig. 2-21, where a constant voltage is maintained across the charging resistor by a zener diode, and an emitter-follower transistor amplifier stage is used. Providing a constant capacitor-charging current over the complete cycle, the circuit has the advantage of economical operation by making double use of the transistor—as a driver for the bootstrap circuit and as an output amplifier. Since the 3.9K load resistor is returned to the negative bus, clipping is eliminated at the bottom of the sawtooth. Transistor Q3 maintains a constant zener current for better linearity and aids Q2 in supplying load current. The linear sawtooth capability of the circuit exceeds 50 kHz.

In Fig. 2-22 the sawtooth generator utilizes a capacitor in place of the zener, a change which eliminates the negative supply. An NPN transistor acts as an output buffer amplifier, with the bootstrap circuit using capacitor C2 and resistor R2 to ensure better sawtooth linearity. The integrating network formed by C1 and R1 provides second-order compensation for any non-linearity. The output waveform can be concaved downward, concaved upward, or linear, simply by adjusting the value of R1. Because the feedback networks are frequency sensitive, capacitor C2 is ineffective at low frequencies and the effective emitter capacity affects linearity and operation as well at high frequencies; therefore, C2 must be very close to .01 mfd. However, by using a higher frequency type UJT (a 2N2647 for instance) the value of C1 could be lowered to .001 mfd or so.

PRECISION TIMING CIRCUITS

Time-Delay Relay

Another useful circuit utilitzing the unijunction transistor, Fig. 2-23, offers a delay in closing a sensitive relay. By closing SW1, capacitor Ct charges to the peak voltage, causing the

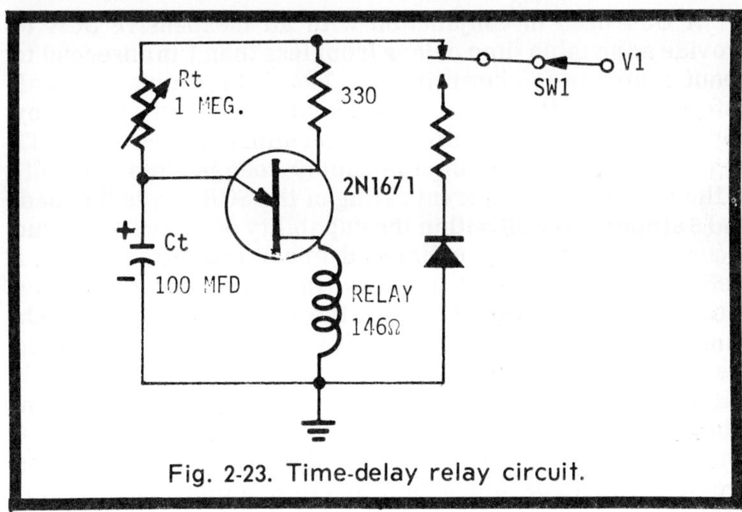

Fig. 2-23. Time-delay relay circuit.

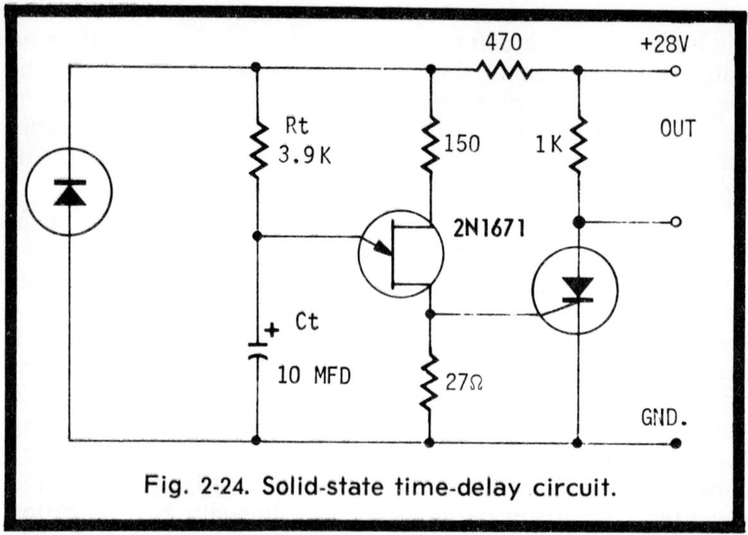

Fig. 2-24. Solid-state time-delay circuit.

unijunction to trigger; the capacitor discharges through the relay coil, closing the hold contacts as well as the external control contacts which may be used for any function desired. For satisfactory operation in a circuit of this type, the relay must have low power requirements, a fast operating time, and a reasonably low coil resistance. The delay is not controlled in any significant manner by temperature or supply voltage, but rather by resistor Rt. The approximate time delay is calculated at 1 second for each 10K in Rt.

Precision Time-Delay Circuits

A UJT used in conjunction with an inexpensive SCR to provide adjustable time delays from less than 1 millisecond to about 1 minute is shown in Fig. 2-24. This timing interval, determined by the value selected for RtCt and started by applying power to the circuit, is terminated when the UJT triggers the SCR, applying full supply voltage (less one volt) to the load. Only the current rating of the SCR limits the load, and 6 amperes is well within the capability of a C20F with stud mounting. A precision calibrated linear resistor, such as a Helipot with 0.5 percent linearity, in place of Rt provides a means of accurately controlling the time delay over a wide range after a single calibration. It is essential for the timing resistor to be small enough to supply the minimum UJT trigger current, Ip, plus the timing capacitor leakage current when the UJT emitter is biased at its peak point voltage. A limit of 3 megohms for Rt is set by the 2N1671's maximum Ip requirement of 2 ua, and with Ct at 4 mfd, a time delay up to 12 seconds is normal.

50

The circuit in Fig. 2-25 permits timing from a fraction of a millisecond up to approximately 5 hours. R1 and C1 control the time delay interval, which is started by applying power, and the controlled rectifier is triggered by the UJT (2N494C). Various load currents are possible, limited only by the capability of the particular SCR used. The value of R1 must be small enough to provide the minimum trigger current for the 2N494C, considering the capacitor leakage current when the unijunction emitter is biased to the peak point voltage. R1 at 3 megohms and C1 at 2 mfd provide an interval of 6 seconds without the 2N491 oscillator. By reducing the minimum Ip requirements on the order of 1000 times, the 2N494C could be pulsed at the upper-base junction with a 750-millivolt negative pulse. Although not too critical, it would be best to use a pulse period less than 0.02 (R1 x C1). The peak point voltage is forced to drop slightly with the negative pulse, and the unijunction transistor will trigger with the necessary Ip supplied from C1, provided the voltage level is greater than this.

To meet the low leakage requirements, C1 should be a mylar capacitor. The value of R2 which provides the best temperature stabilization over the desired operating range can readily be ascertained. If the timing circuit must be isolated from the controlled rectifier circuit (which possibly could be common to the AC line), the 27-ohm resistor should be removed and a pulse transformer substituted in its place.

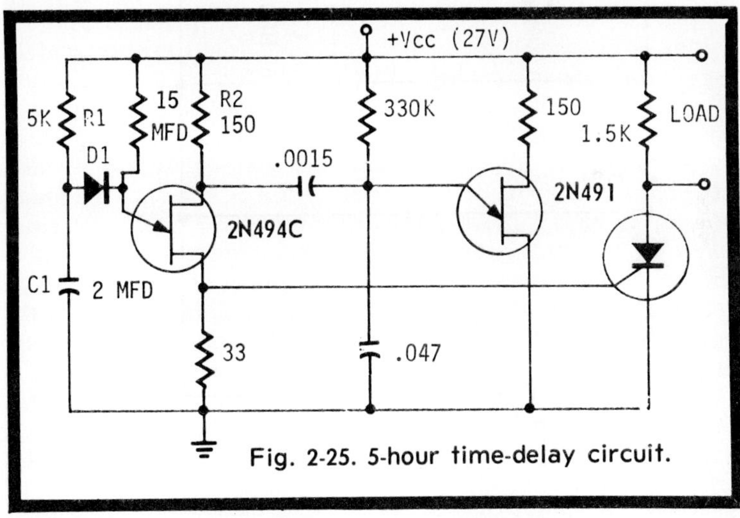

Fig. 2-25. 5-hour time-delay circuit.

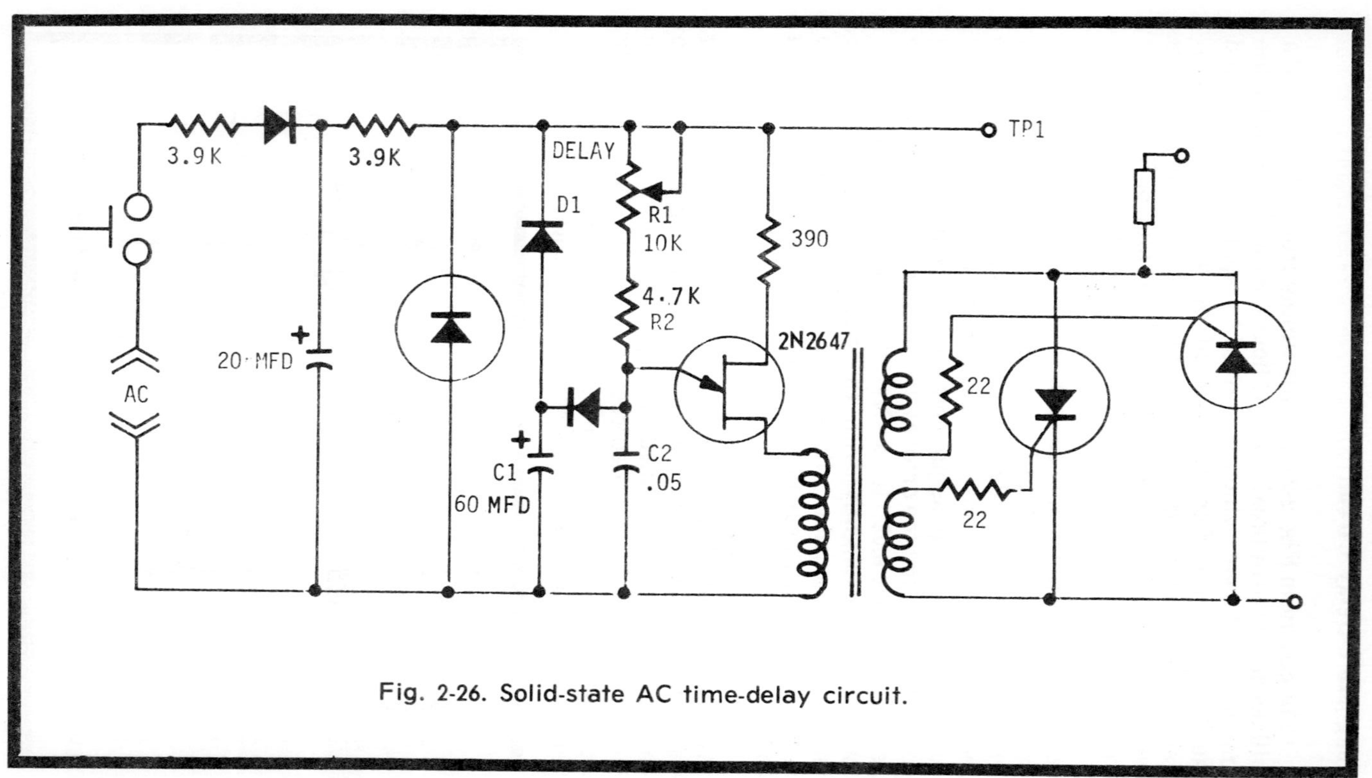

Fig. 2-26. Solid-state AC time-delay circuit.

Before triggering, the 2N494C UJT has an input impedance greater than 1500 megohms, and the maximum obtainable time delay is governed by maximum R1-C1 values possible with the low leakage requirement. Minus diode D1, the R1 limit is 15 megohms for an accuracy of 5 percent at 55 degrees C. However, the use of the diode permits R1 values up to 10 kilomegohms.

AC Time Delay

A time-delay circuit with an AC output is shown in Fig. 2-26. Using only one UJT, the circuit is started by closing the power switch to activate the unijunction circuit. As the voltage at the unijunction emitter arrives at the peak point voltage, the UJT oscillates at a high frequency, governed by the time constant (R1 + R2)C2, with C1 remaining charged. The UJT output is coupled through the pulse transformer and turns the SCRs on to apply voltage to the load. By oscillating at a much higher frequency than the regular line frequency, the UJT readily switches the silicon controlled rectifiers from full on to full off. Since a high output pulse is needed to fire a pair of SCRs in parallel, the 2N2647 is used. With the starting switch in the open position, a discharge path for C1 is provided by diode D1. Larger time delay is available simply by increasing C1's value.

TUNNEL DIODE OSCILLATORS

As an oscillator the tunnel diode offers a number of features: high-frequency capability, low power consumption,

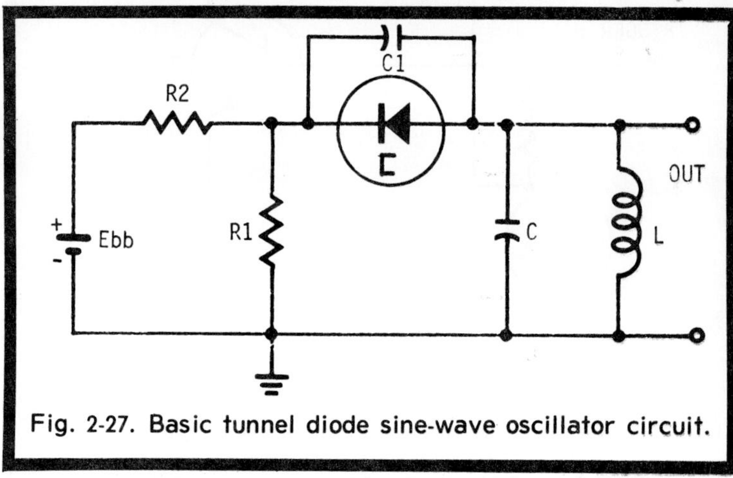

Fig. 2-27. Basic tunnel diode sine-wave oscillator circuit.

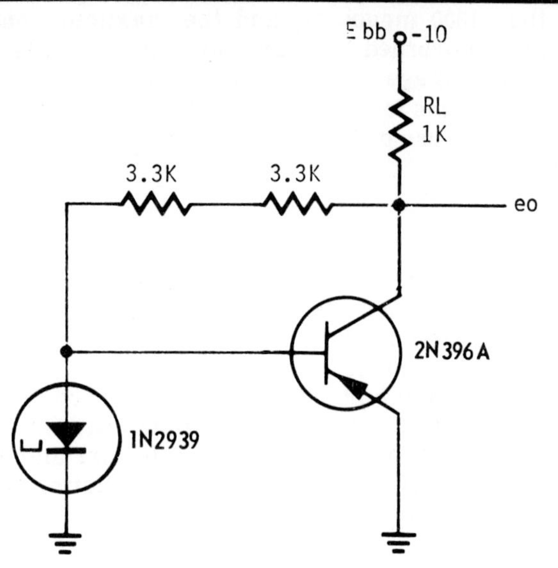

Fig. 2-28. Astable hybrid oscillator circuit.

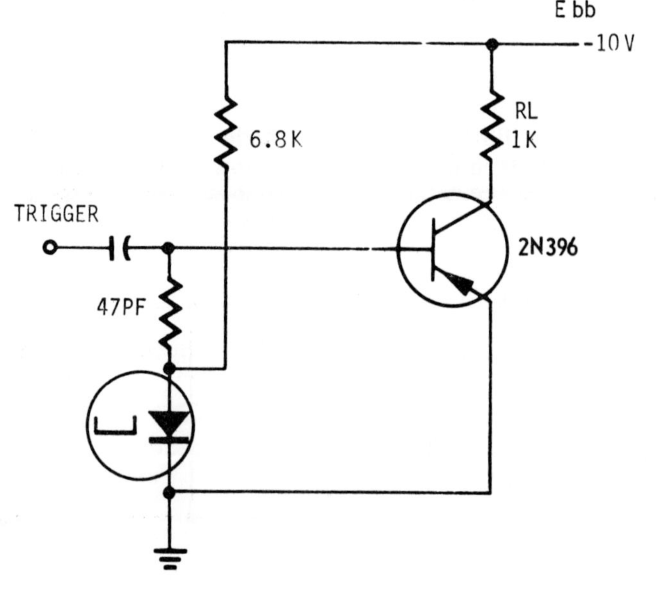

Fig. 2-29. Bistable tunnel diode circuit.

good frequency stability, and extreme circuit simplicity–
advantages that enable a designer to produce a stable
miniature oscillator circuit with a wide range of uses. Fig. 2-27
illustrates the basic series-parallel sine-wave oscillator. The
low noise figure of the tunnel diode also adds to its attraction
as an oscillator.

A major difference between the usual diode and the tunnel
lies in the high conductivity of the latter–a thousand times the
conductivity of the conventional diode. The concentration of
impurities while forming the diode produces a narrow junction
which permits the electrons to tunnel under the barrier. When
operating as an oscillator an ideal property is the negative
resistance at low forward-bias values. Functioning to
replenish losses in the tank circuit, oscillations are sustained.
When bias across the diode decreases, it supplies more
current as a result of its negative-resistance characteristic.

TUNNEL DIODE MULTIVIBRATORS

Having exceptionally high-speed switching capabilities
and stable characteristics, the tunnel diode has proven useful
in many multivibrator applications. As the fastest switching
device known, speeds are normally limited by the package
inductance and capacitance of the circuit rather than the
diode.

An example of the astable tunnel diode circuit is shown in
Fig. 2-28. The transistor amplifier raises the low-level output
of the generator to a reasonable amplitude without undue
loading. With very few components and low power
requirements, a fine quality square wave is offered by this
hybrid circuit.

In the bistable multivibrator circuit in Fig. 2-29, the tunnel
diode is used with a PNP germanium transistor amplifier to
produce a narrow pulse when triggered. Excellent results are
realized with a minimum number of parts in this type of
generator.

The very unusual staircase-wave generator or pulse
frequency divider is illustrated in Fig. 2-30. Tunnel diodes are
used in a series connection with the transistor acting as a
shorting-resetting element. Although there are limitations in
the use of two-terminal devices such as the tunnel diode,
wherever the desired reliability can be designed in the circuit,
their inherent characteristics should not be overlooked.

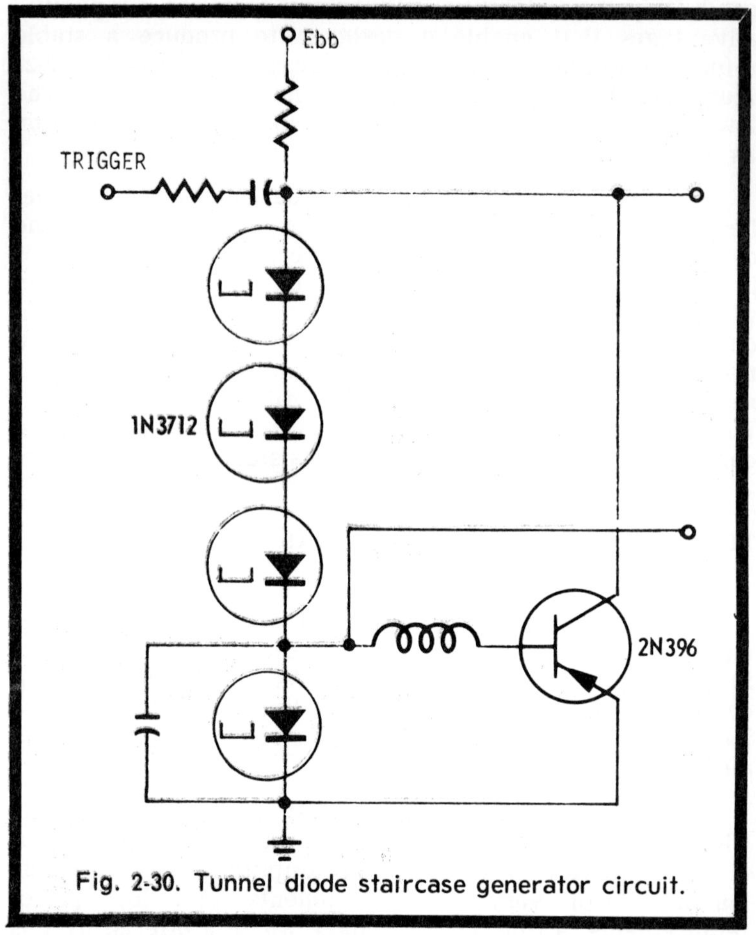

Fig. 2-30. Tunnel diode staircase generator circuit.

IC PULSE GENERATORS

The HEP-581 (4-input gate) can be used to convert the sine-wave output from any available source having a peak-to-peak amplitude of 2 volts or more to a square wave which may be adjusted to any desired frequency below 30 kHz. Fig. 2-31 shows the circuit used; the 100K potentiometer is used to adjust the frequency and the 5K pot, the output level. An electrolytic capacitor couples the sine wave to the base of the input transistor at terminal 1. As the gate is biased from saturation to cut-off by the incoming wave, the inverted square wave is produced, inverted again, and then amplified before leaving the IC. The TO-99 packaged device is the equivalent of five transistors.

56

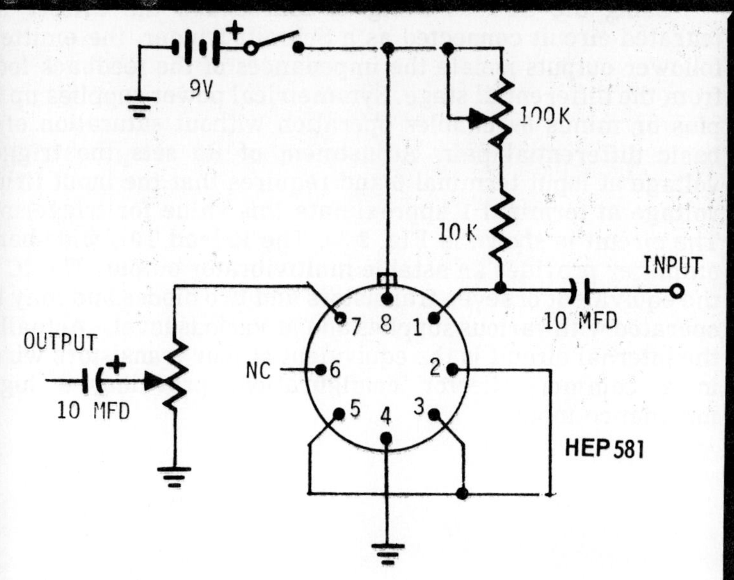

Fig. 2-31. Integrated circuit square-wave converter.

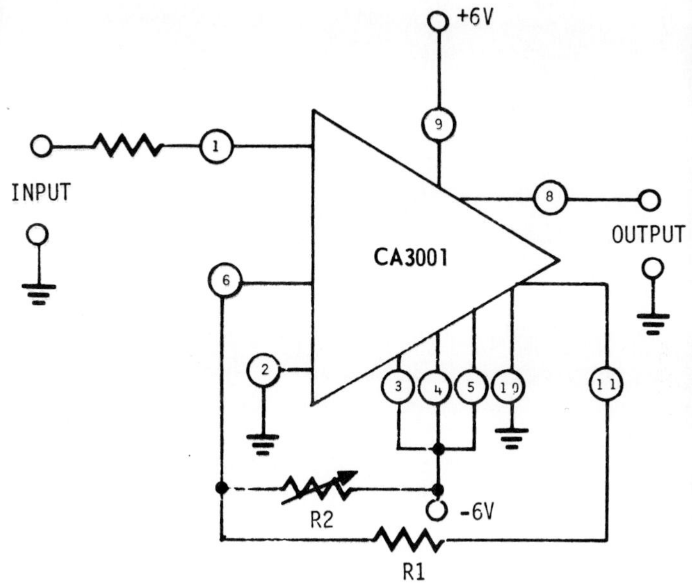

Fig. 2-32. IC square-wave generator diagram showing outboard components and connections.

Using the UJT as a signal source and the CA3001 integrated-circuit connected as a Schmitt trigger, the emitter-follower outputs isolate the impedances of the feedback loop from the differential stage. Symmetrical power supplies up to plus or minus 6v enables operation without saturation of a basic differential pair. Adjustment of R2 sets the trigger voltage at input terminal 6 and requires that the input firing voltage at terminal 1 approximate this value for triggering. The circuit is shown in Fig. 2-32. The 12-lead T0-5 wide-band amplifier provides an astable multivibrator output. The IC is the equivalent of seven transistors and two diodes and may be operated with various supplies and at various levels. Actually, the internal circuit is the equivalent of four transistors wired in a common-collector configuration, providing a high-impedance input.

Chapter 3

Response Characteristics

Analysis of sine-wave circuits is a basic operation due to the comparative ease of such measurements, plus the fact that a sine wave usually survives in its native form after passing through any combination of R, L, and C. However, the non-sinusoidal wave in passing through most any such network will come out with an entirely different shape that would seldom, if ever, bear any likeness to the original. As the very shape of the pulse wave will change in going through an RC, RLC, or RL network, the normal relationship between voltage and current can no longer be formulated in terms of capacitive reactance, inductive reactance, and impedance. The Ohm's Law formula (I equals E divided by Z) that provides the relationship between voltage and current applies only to sinusoidal waves because they are not distorted in passing through RLC circuits.

An RLC, RC, or RL circuit may change the level of a sine wave, or even alter the phase to some degree, but the output nevertheless still will be regarded as a sine wave. In the event that a nonsinusoidal wave should be applied to a circuit without inductance or capacity, the wave will not change in shape as a result of passing through a pure resistance type circuit. Applying a square wave to an RLC, RC, or RL circuit could provide an output having the appearance of a sawtooth wave. We may rest assured that the original square-wave input will be distorted to a noticeable degree in shape at the output, as clearly shown in Fig. 3-1.

Any periodic nonsinusoidal wave may be assumed to be made up of a number of superimposed pure sine waves, with the lowest frequency component representing the fundamental. Since all the remaining sine-wave components are harmonics in the case of the nonsinusoidal wave, they bear an exact multiple-frequency relationship to the fundamental.

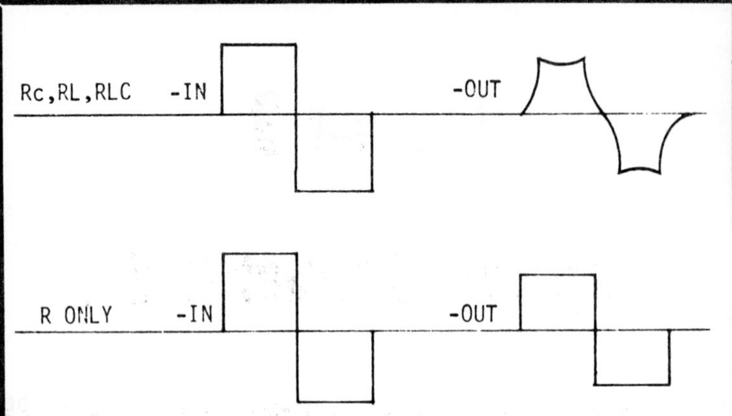

Fig. 3-1. Waveforms illustrating typical response characteristics of linear circuits.

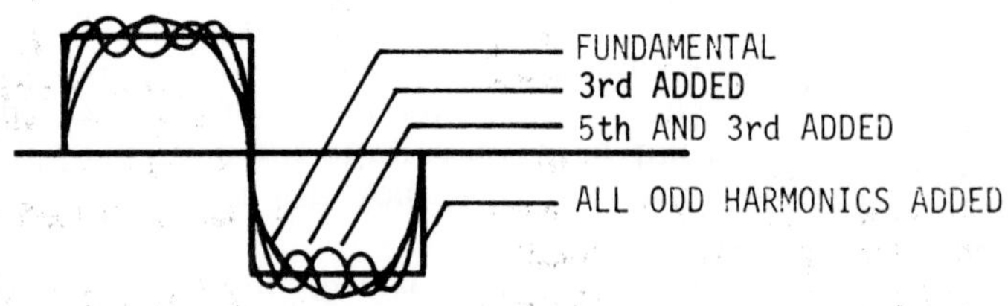

Fig. 3-2. Composition of a square wave.

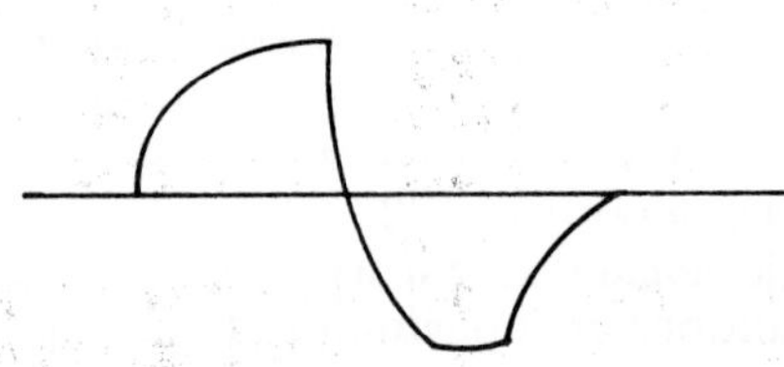

Fig. 3-3. Square wave showing effects of poor high-frequency
response.

A square wave is composed of an infinite number of odd harmonics as shown in Fig. 3-2. It is possible to analyze the response of a linear circuit to a nonsinusoidal wave by evaluating the response of such a circuit to each of the sine-wave components, since they are not distorted by linear circuits. This is readily understandable as all linear circuits are made up of capacitors, resistors, and inductors. Square waves, sawtooth waves, and numerous other nonsinusoidal waves of greatest interest in pulse work, unfortunately contain many sine-wave components of sufficient amplitude to demand consideration, thus making this method of analysis quite tedious. As a result, the sine-wave method is seldom used in pulse work, but dividing the nonsinusoidal wave into sine-wave components provides a useful tool in studying the causes and formulating a means of correction for waveform distortion.

Normally, when harmonics increase in order, they decrease in amplitude, so that all harmonics above a certain order may be ignored and any approximation to the original waveform will still be quite acceptable. How satisfactory such a method is depends on the number of harmonics included and how many were passed up. The closer the analysis is held to a pure sine wave the greater the amplitude drop of the harmonics. Therefore, a square-wave pulse with sharp corners does contain many high-order harmonics which could not be ignored if sharp corners are required in the output. A rounded wave or pulse is much more like a sine wave, thus high-order harmonics can be overlooked without much change in waveshape simply because their amplitudes drop off so much faster than the square pulse. If a square wave is applied to a circuit with poor high-frequency response, the wave is distorted as shown in Fig. 3-3, as the necessary high frequencies are not reproduced.

Once a nonsinusoidal wave has been divided into its sinusoidal components, which in itself is a mathematical task beyond the scope of this discussion, the response of a linear circuit to the wave may be obtained by applying the ordinary circuit reactance and impedance laws used for sine waves to each component of the input voltage. Using the sine-wave laws, the output amplitude and phase are calculated for each component of the input voltage. Obviously the circuit will cause amplitude changes and phase shifts that are different for each frequency, so that when the individual components are again added at the output, the result is the actual distorted

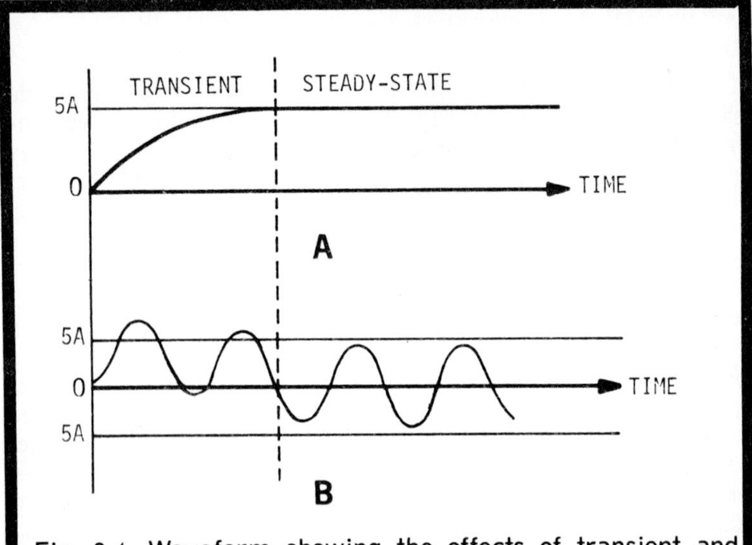

Fig. 3-4. Waveform showing the effects of transient and steady-state response.

output wave. As already mentioned, such an analysis (harmonic analysis), while being theoretically sound, is difficult to carry out. A different analysis method, much more suitable for pulses, is covered later in this chapter.

TRANSIENT AND STEADY-STATE RESPONSE

Two types of response may occur when a voltage is applied to a circuit—transient and steady-state. Here the word response refers to how the input current varies with time, or, if the circuit has an output, how the output voltage varies with time. The current or voltage variation that occurs for a short time immediately after a sudden change in the applied voltage, as when a switch is closed, is called transient response. The current or voltage wave that appears in the circuit sometime later, after the effects of the sudden change in applied voltage have worn off, is called the steady-state response.

Fig. 3-4A shows the transient and steady-state response conditions that occur after switch S is closed in the RL circuit shown in Fig. 3-8. Because of the circuit inductance the current is zero the moment the switch is closed, as the graph shows, but it builds up fast at first and slower later until the steady-state value of 5A is reached. In Fig. 3-4B a sinusoidal

62

AC voltage is applied to an RL circuit. After the switch has been closed a few cycles, the circuit current becomes a steady AC sine wave of constant amplitude as shown. But when the switch is first closed, the wave varies in amplitude, cycle duration, and shape.

Any sudden change in circuit conditions will cause a transient response. For example, if a square-wave pulse is applied to a circuit, a transient response occurs. That is why transient response is of great interest in the study of pulses. Generally, a transient response is aperiodic, since the circuit disturbance that causes the transient is not usually repeated at regular intervals. We have previously stated that a sinusoidal waveform will not be distorted in shape by a linear circuit. To be precise, we must say that the steady-state response of a linear circuit to a sine-wave input is always sinusoidal. The transient response is not necessarily sinusoidal, as Fig. 3-4 shows.

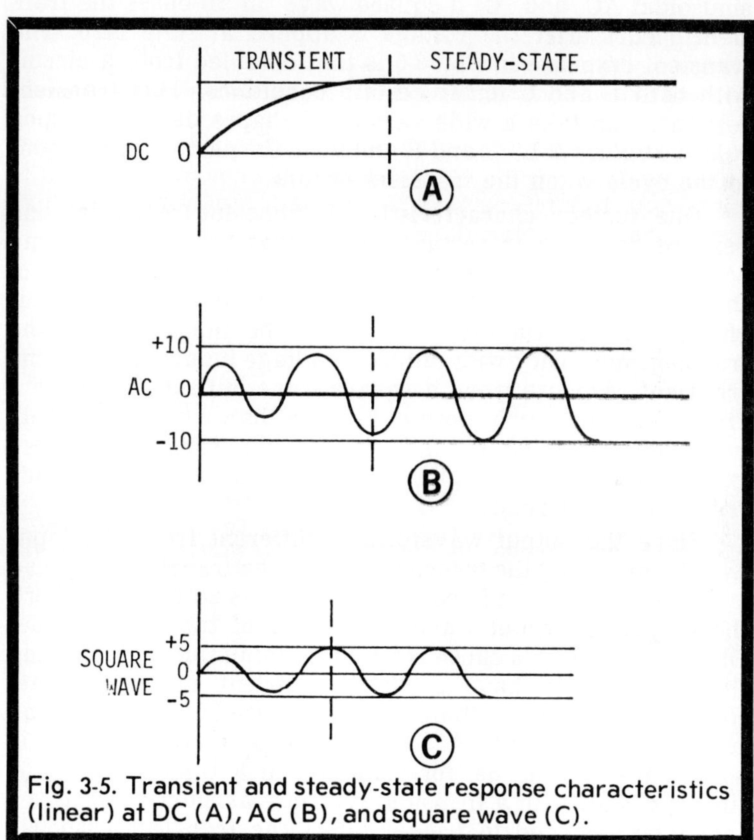

Fig. 3-5. Transient and steady-state response characteristics (linear) at DC (A), AC (B), and square wave (C).

The steady-state response of any circuit consists of a periodic wave plus a DC component, either of which may or may not be zero. For example, the steady-state response in Fig. 3-4A has a DC component of 5A and no AC (periodic) component. The steady-state response of Fig. 3-4B has a periodic component of 5A peak value, but no DC component. The unfiltered output from a full-wave rectifier is an example of a steady-state response with both a periodic and a DC component. If a linear circuit is completely passive - that is, contains no batteries or other power sources - the DC component of any output voltage cannot be greater in magnitude than the DC component of the input voltage. Furthermore, the periodic component of the steady-state response must have exactly the same frequency as the input voltage.

Fig. 3-5 indicates transient and steady response characteristics of a hypothetical linear circuit to (A) DC, (B) sinusoidal AC, and (C) a square wave. In all cases the transient occurs after the voltage is applied at time zero. The transient response in Fig. 3-5 is that obtained from a circuit with both L and C under certain conditions. This transient response can take a wide variety of shapes depending upon relative values of L, C, and R and upon the particular moment in the cycle when the transient occurs.

One further characteristic of transient response, and perhaps the most important one, is that the length of time occupied by the transient after a disturbance is independent of the nature of the input voltage and depends only on circuit characteristics. On the other hand, the magnitude of the transient may vary with different voltage inputs. That is, the transient response when a square wave is first applied to a circuit may be much more noticeable than when a sawtooth wave is first applied, but in both cases the transient response periods are equal in length, since this characteristic depends only upon the circuit.

Since the output waveform is different from the input waveform during the transient period, the transient response is a form of distortion in equipment, such as amplifiers, where the output and input waves should be of the same shape. Distortion from this cause is held to a minimum by designing the circuitry to have a short transient response time. Response time of less than one microsecond is found in some practical equipment. A long transient response time is sometimes needed, as, for example, in a timing circuit. A timing circuit with a transient response as shown in Fig. 3-5 might be designed to trigger when the transient response value is three-fourths the steady-state value. Then, by adjusting the transient response time, the timer can be made to trigger at the desired time. The transient response time is determined by the time constant of the circuit, which is discussed later.

AMPLIFIER RESPONSE TO NONSINUSOIDAL WAVEFORMS

We have seen that any RLC, RL, or RC circuit will distort a nonsinusoidal waveform while it passes from input to output. Much electronic equipment, however, must pass non-sinusoidal waveforms, or, more specifically, amplify non-sinusoidal waveforms or pulses of voltage without distortion. In terms of the sinusoidal-components approach, this means that the amplifier must be capable of amplifying all frequency components—the fundamental and all harmonics—equally. Furthermore, either there must be no phase shift or else all frequencies must be shifted in time by an equal amount.

"No phase shift" means that the signal appears at the output of the amplifier at precisely the same instant as it appears at the input, a condition seldom obtained in practice. However, as long as all parts of the signal (that is, all frequency components) are delayed by the same number of microseconds, all parts of the output signal still have the same relationship to each other and no distortion due to phase shift. This is called linear phase shift. Since the duration of one cycle is inversely proportional to frequency, in order to obtain a linear phase relationship the phase shift in degrees must be proportional to the frequency.

For example, if an amplifier causes the phase of a 1000-Hz signal to lag by 10 degrees, then in order to avoid phase-shift distortion the output of a 2000-Hz component must lag the input by 20 degrees, a 3000-Hz component by 30 degrees, and so on. With equal amplification at all frequencies, and a linear phase-shift characteristic, the output waveform will be exactly the same shape as the input but will be delayed slightly in time. That is, the output pulse will come out of the amplifier slightly after the input pulse goes in.

In many circuits the delay caused by phase shift is so small that it can be neglected. However, in timing circuits (radar, for example) accuracy depends upon knowing how much time elapses between the time the signal is received at the antenna and the time it is actually recorded. A perfect amplifier which amplifies all frequencies equally and introduces a linear phase shift is, in practice, impossible to build. However, extremely good signal reproducibility can be obtained with wideband amplifiers that equally amplify, within reason, the fundamental sinusoidal component and the harmonics that cannot be considered to be negligible in amplitude and that exhibit fairly linear phase shifts over the desired frequency range. Naturally, the wider the frequency response, the greater the number of harmonics that will be amplified and the more closely the output signal will resemble the input signal.

Such wideband amplifiers require special circuitry; ordinary AF and RF amplifiers by no means fulfill the requirements. For example, since the ear is relatively insensitive to phase shift, AF amplifiers are usually designed only for good frequency response. Even then the response need be no wider than the response of the human ear, since the higher frequencies cannot be heard anyhow. In ordinary AM radio RF and IF amplifiers, phase shift is not important; furthermore, the selectivity of the unit depends on its having a relatively narrow bandwidth, making it very unsuitable for amplifying nonsinusoidal waves or pulses.

As an example of how poor frequency response can distort a wave, consider Fig. 3-3. The waveform illustrates what happens to an ideal square wave which is fed to an audio amplifier—the resulting distortion due to poor low-frequency response, or a reduction in amplification at the lowest frequency or fundamental, and the resulting distortion due to poor high-frequency response, or a reduction in amplification at harmonics which are not negligible in amplitude.

In addition to the distortion of a wave introduced by poor frequency response and nonlinear phase shift, another type of distortion is introduced by tubes and transistors since they are not completely linear devices. In other words, even a perfect sine wave at the input of a tube or transistor circuit will not come out a perfect sine wave. Such distortion is termed harmonic distortion. The distorted waveshape is made up of a pure, fundamental sine wave plus harmonics of that fundamental. Thus, the nonlinear amplification characteristics of tubes and transistors result in the generation of spurious frequencies not in the original signal. Satisfactory amplification of square pulses requires a high-quality wide-band amplifier. All the types of distortion mentioned must be held to a minimum for satisfactory pulse reproduction.

PULSE PARAMETERS

Rectangular pulses are very widely used. Ideally, they should have vertical sides; that is, the voltage should jump

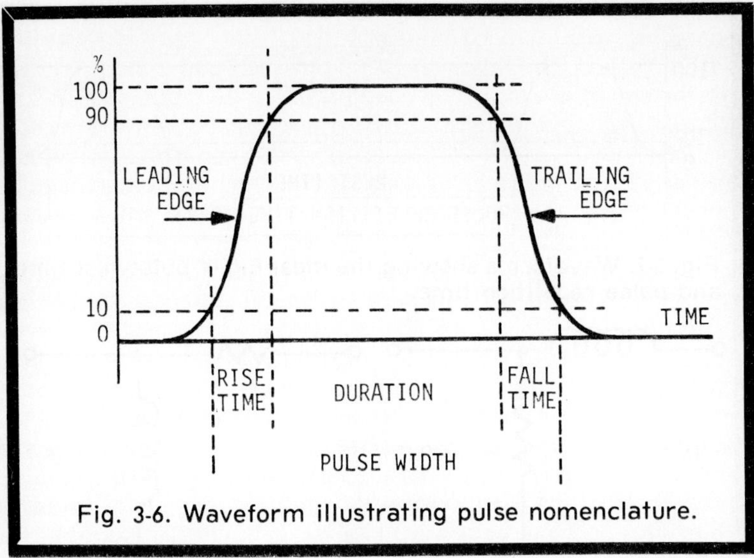

Fig. 3-6. Waveform illustrating pulse nomenclature.

from zero to full pulse value instantly and stay at full value for the desired pulse duration. But, practically, any rectangular pulse must have sloping sides because of the time required for a pulse to form and decay. The time required for a pulse to jump from 10 up to 90 percent of its maximum value is called pulse rise time (Fig. 3-6). Similarly, the time required for a pulse to fall from 90 down to 10 percent of its maximum value is called the pulse decay or fall time, as explained in Chapter 1. The reason we use 10 and 90 percent rather than 0 and 100 percent in measuring rise and fall times is that, as Fig. 3-6 shows, the rounded top and bottom corners of the leading and trailing edge make it hard to tell exactly when the rise and fall start and end.

The shorter the rise and fall times of a pulse, the higher the frequency response of amplifiers and other pulse-handling circuitry must be. The time interval between the end of the rise time and the beginning of the fall time is called the pulse duration. The pulse width is the sum of the rise, duration, and fall times. The time between successive pulses is called the rest time (shown in Fig. 3-7). For periodically occurring pulses, the time between a point on one pulse and the corresponding point on an adjacent pulse is called pulse repetition time (PRT) or pulse period. The PRT should not be confused with the pulse repetition frequency (PRF). The PRF is the number of pulses that occur per second. The PRF is also called the pulse repetition rate (PRR). The pulse repetition

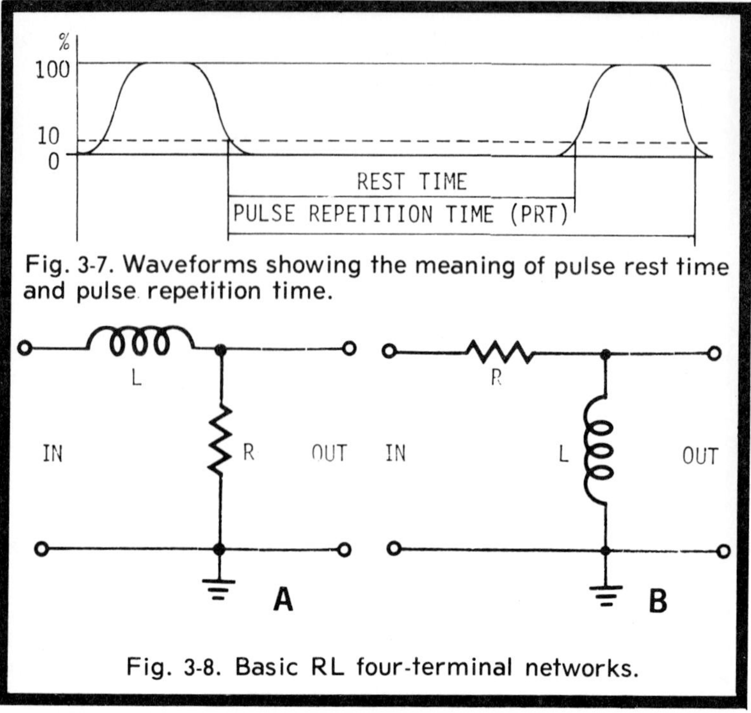

Fig. 3-7. Waveforms showing the meaning of pulse rest time and pulse repetition time.

Fig. 3-8. Basic RL four-terminal networks.

rate is equal to the reciprocal of the pulse repetition time, and conversely, the pulse period equals the reciprocal of the pulse repetition frequency.

In practical circuitry we classify a pulse as rectangular if the rise and decay times are short compared to its duration time. For example, many such pulses have rise and decay times of a microsecond or less and duration times of the order of a millisecond. For all practical purposes the pulse is rectangular, although a true rectangular pulse would have zero rise and decay times.

FOUR-TERMINAL NETWORKS

Until now we have been considering the response of a circuit as being the current that flows when a step voltage is applied, or vice versa. This, in fact, is the only possible response to consider with a one-element circuit. With two elements in the circuit, however, we can make an input-output type, or 4-terminal network. The 4-terminal network is perhaps the most useful and most often encountered arrangement. With such a circuit, we want to know the output

68

voltage or response when a step voltage is applied at the input.

A simple 2-element series RL circuit can be connected to form two such 4-terminal networks, as Fig. 3-8 illustrates. In the first (A), the output voltage appears across the resistor and the second (B) across the inductor. The response, or voltage present, when the output is taken across the resistor as in Fig. 3-8, is, by Ohm's Law, simply the current that flows in the circuit–the same current we have been discussing as given by the current-response curve–times the value of the resistance.

Fig. 3-9 illustrates this fact. Notice that the maximum or steady-state output voltage equals the input. In other words, when the steady state is reached, the inductor appears as a short circuit. At one time constant, the current is 1.26A. Therefore, the voltage at this moment is 1.26 X 16 or 20.1v. If we consider the voltage across the inductor in Fig. 3-9, we find that it is equal to the voltage across the resistor subtracted from the input voltage. This is apparent when we consider Kirchhoff's Voltage Law, which states that the sum of the voltages across all the passive elements in a series circuit must equal the applied voltage.

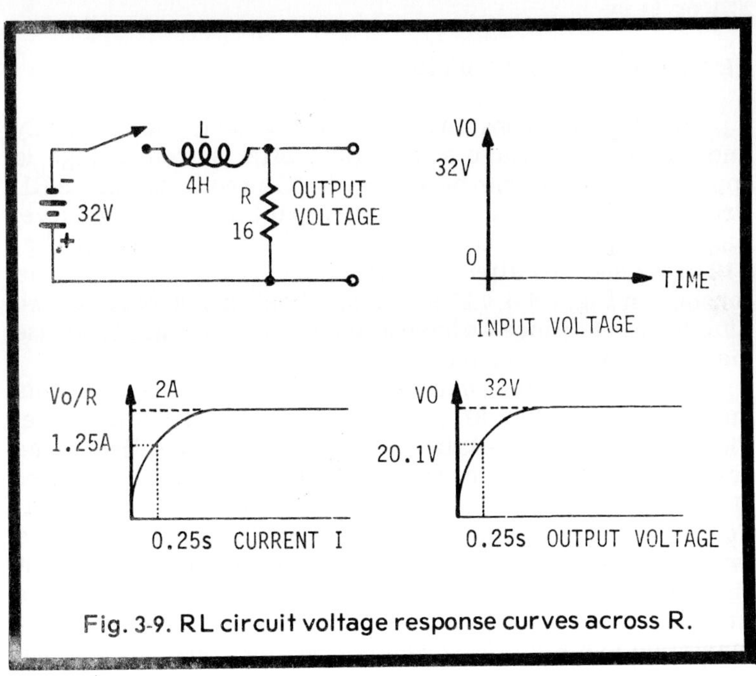

Fig. 3-9. RL circuit voltage response curves across R.

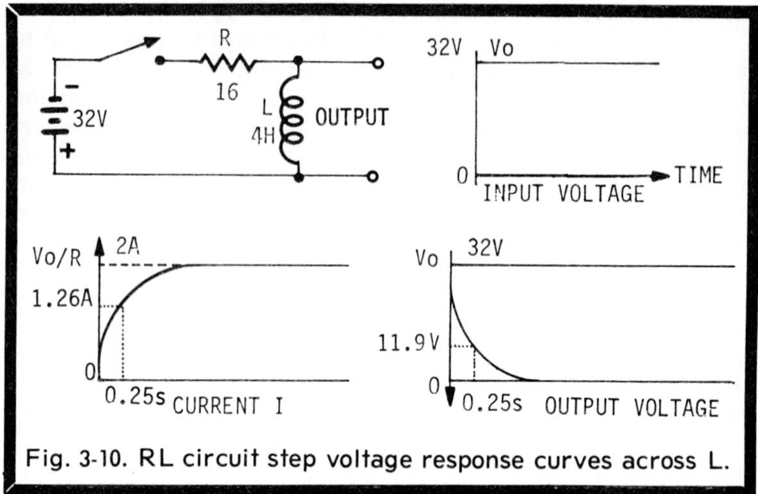

Fig. 3-10. RL circuit step voltage response curves across L.

Fig. 3-10 shows the circuit if we take the output across the inductor. Notice that the output at each moment is simply the input less the voltage across the resistor. For example, the voltage across the resistor is, from Fig. 3-9, 20.1v at 0.25 seconds. Hence, the voltage in Fig. 3-10 at 0.25 seconds is 32 - 20.1 or 11.9v.

TIME CONSTANT CURVES

With the time constant curves in Fig. 3-11, you can easily find current or voltage at any time after a step voltage is applied. The curves can be used with RC circuits as well as RL circuits. To see how to use the universal time-constant curve, suppose you want to find the value of the output voltage in Fig. 3-12 at 0.5 seconds after the switch is closed. Since one time constant in Fig. 3-9 is 0.25 seconds, 0.5 seconds is equal to two time constants. Since we have a rising curve in Fig. 3-9, we use the rising Curve A of the time-constant chart.

Reading from the chart, we see that at two time constants the voltage reaches 87 percent of its steady-state value. Therefore, the steady-state value in Fig. 3-9 is 32v. Hence, the voltage 0.5 seconds after the switch is closed is 0.87 X 32 or 27.8v. The current at this moment would be 0.87 X 2 or 1.74A. To find the output voltage in Fig. 3-10 at 0.5 seconds after the switch is closed, we use Curve B of the universal time-constant chart, since a decaying voltage is involved. From Curve B we find that at two time constants the voltage is 13 percent of its maximum value. 0.13 X 32 equals 4.16v, the voltage 0.5 second after the step voltage was applied at the input by closing the switch. Notice that the values on Curve B in Fig. 3-11 are the values of Curve A subtracted from 100. Hence, only one of the curves of Fig. 3-11 is essential in order to work out transient voltage or current values.

SERIES RC AND RL RESPONSE

When we discussed the response of a pure capacitance, we used a constant-current generator to produce a step current and considered the voltage as the response. We could as well examine the response of an RC circuit to a current step; however, since most practical power sources, such as batteries and generators, appear to be more like constant-voltage sources, we shall continue in the study of RC circuits to use step-voltage excitation, with current or output voltage as the response. The reason we excited the pure capacitance with a step current rather than a step voltage was to prevent the current from becoming infinite. This would happen, since an uncharged capacitor, at the instant the step is applied, appears to be a short circuit. However, now that there is a resistance in series with the capacitor, the current is limited to a value calculated by Ohm's Law, Vo divided by R. Thus, in a series RC circuit, a step voltage causes the current to jump immediately to Vo divided by R.

Furthermore, the current begins falling as the capacitor charges, until the current reaches zero. That is, the steady-state current response to a step voltage is zero and the

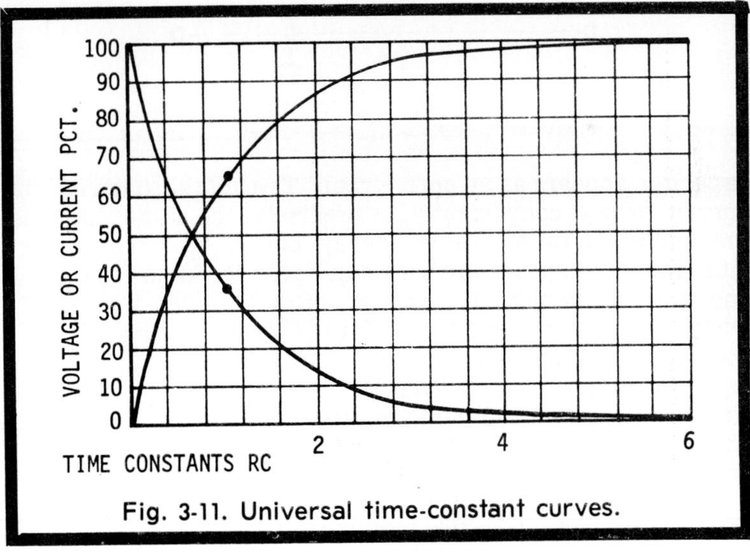

Fig. 3-11. Universal time-constant curves.

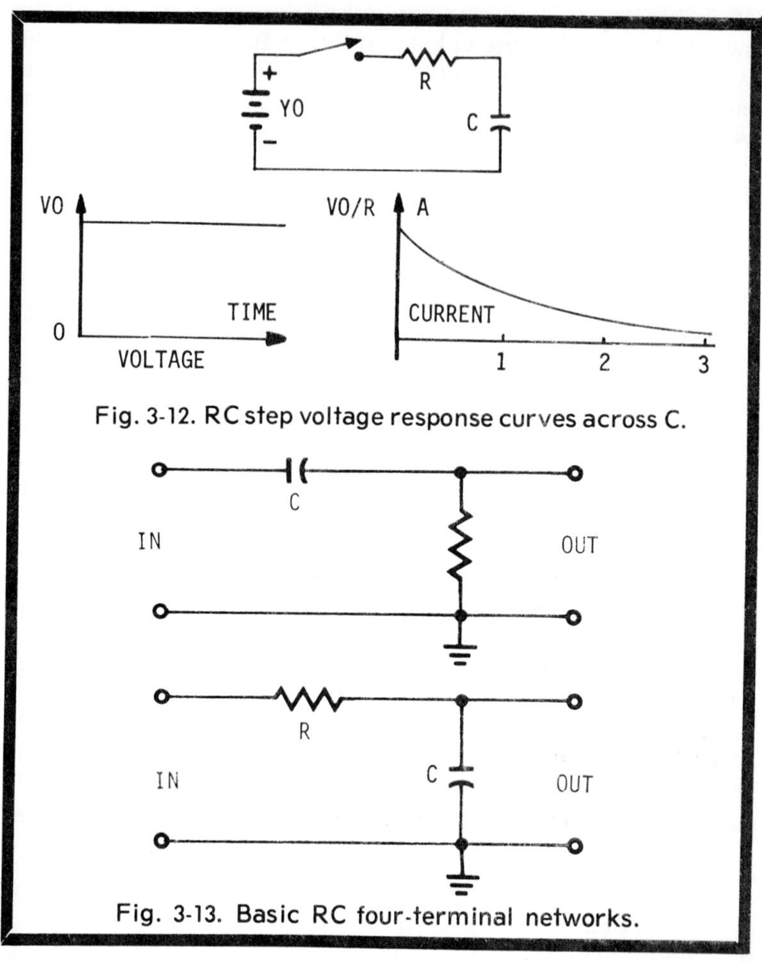

Fig. 3-12. RC step voltage response curves across C.

Fig. 3-13. Basic RC four-terminal networks.

capacitor appears as an open circuit. The actual shape of this
current decay, conveniently, is precisely the same shape as
the universal time-constant decay curve. The circuit, excitation, and response are shown in Fig. 3-12. In an RC circuit
one time constant is the time, in seconds, given by R X C,
where R is in ohms and C is in farads, or more conveniently,
you can use R in megohms and C in microfarads. Thus, in
using the decay curve to find the current response of a series
RC circuit to a step voltage, the 100 percent amplitude
corresponds to Vo divided by R, while one time constant
corresponds to R X C.

As an example, consider a 100v battery connected across a
100K resistor in series with a 1.0-mfd capacitor. The maximum

72

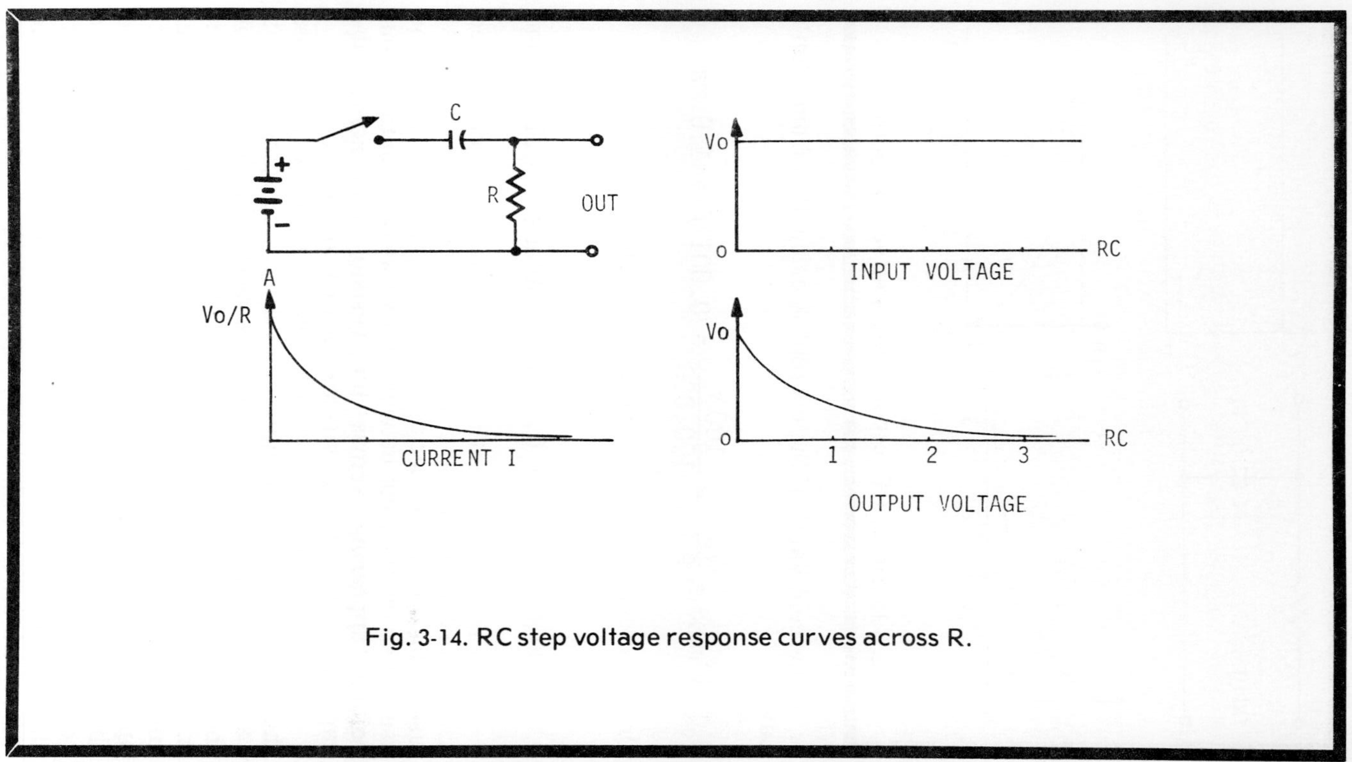

Fig. 3-14. RC step voltage response curves across R.

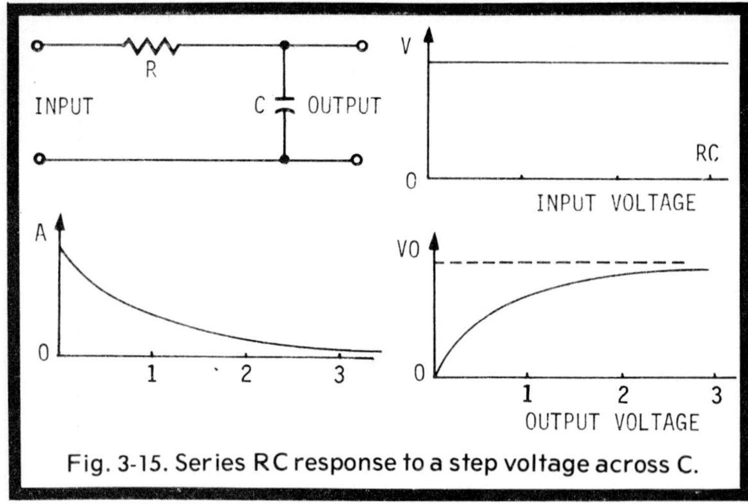

Fig. 3-15. Series RC response to a step voltage across C.

or 100 percent value of the current is calculated from Ohm's Law:

$$I \text{ max} = \frac{Vo}{R} = \frac{100 \text{ v}}{100,000} = 0.001 \text{ A} = 1.0 \text{ ma}$$

One time constant is calculated as follows:

$$t = R \times C$$
$$= 100,000 \text{ ohms} \times 0.000001 \text{ farads} = 0.1 \text{ second}$$

With these two quantities calculated, they can be used easily to scale the decay curve in the manner we have been discussing. As with the series RL circuit, the RC circuit can be used as an input-output network in two ways: with the output voltage appearing across the resistor and across the capacitor. These two circuits are illustrated in Fig. 3-13.

Considering first the circuit that takes the output across the resistor, we find that the output voltage, according to Ohm's Law, is the current times the resistance. In the terms of the universal time-constant decay curve, this means that maximum voltage output equals the input voltage; the time-constant length remains fixed. Fig. 3-14 shows the circuit, applied voltage, current, and output-voltage response. When the output is taken as the voltage across the capacitor, the response curves appear as in Fig. 3-15. As with the RL circuit, the voltage across the capacitor is the applied step, less the voltage across the resistor. This gives a curve with the shape

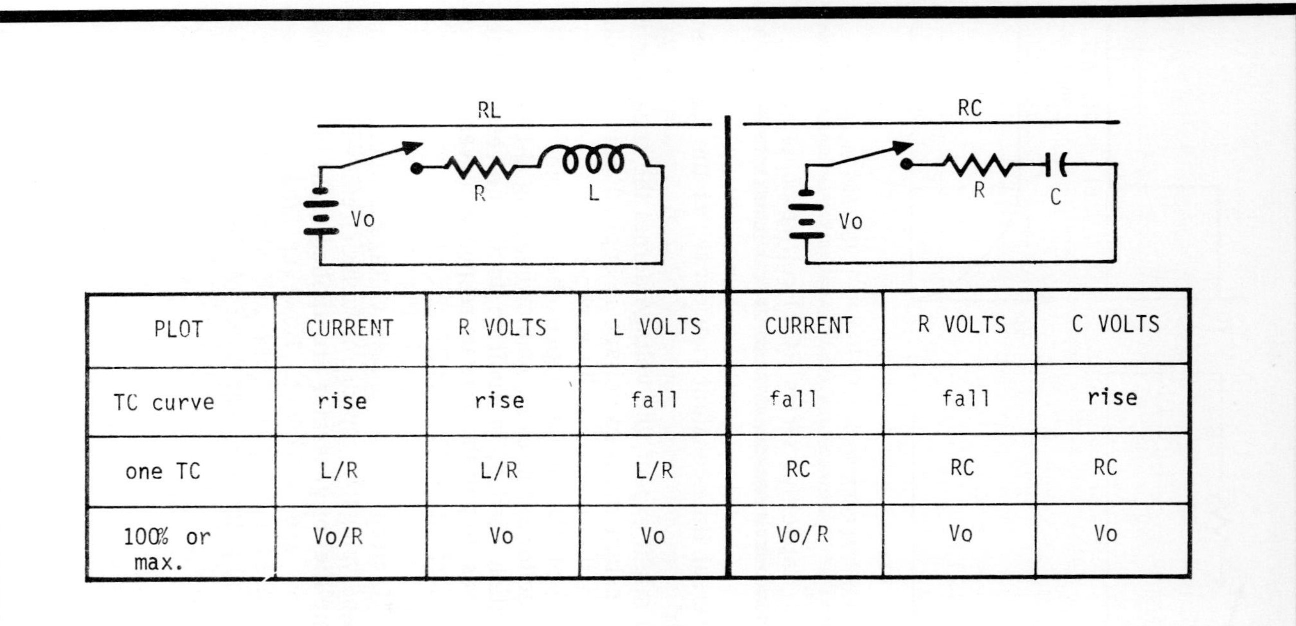

PLOT	CURRENT	R VOLTS	L VOLTS	CURRENT	R VOLTS	C VOLTS
TC curve	rise	rise	fall	fall	fall	rise
one TC	L/R	L/R	L/R	RC	RC	RC
100% or max.	Vo/R	Vo	Vo	Vo/R	Vo	Vo

Fig. 3-16. Response-curve plotting data for step voltage Vo.

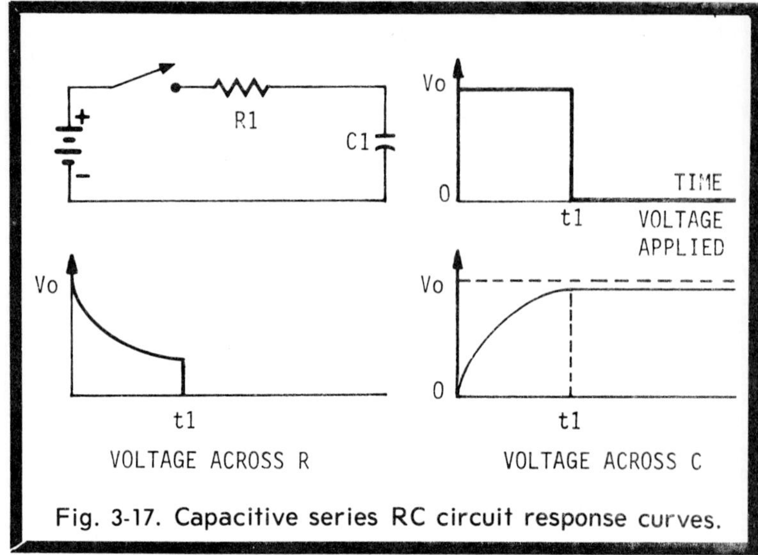

Fig. 3-17. Capacitive series RC circuit response curves.

of the universal time-constant rise curve, as the figure indicates.

As you might expect, the voltage across the capacitor is zero at the instant the step voltage is applied, since the uncharged capacitor is essentially a short circuit. As steady-state is reached, the voltage approaches Vo, since the fully charged capacitor becomes an open circuit. Fig. 3-16 gives all the information needed, in addition to the universal time-constant curves, to plot the various response curves we have been discussing in series RL and RC circuits.

Thus far, we have been considering voltages and currents in RC and RL circuits when a step voltage is applied at the input. For analyzing the response of such a circuit when pulses from an amplifier or generator are applied to the input, these considerations are sufficient. However, many circuits—especially those involving switches—are more easily analyzed from a very slightly different point of view. For example, consider Fig. 3-17. At time 0 the switch is closed, and the voltages begin varying as the universal time-constant curves predict. If, however, at some time later (designated as t1 in the figure) the switch is opened, the applied voltage drops to zero.

As we have stated earlier, this drop to zero can be thought of as the sum of the original positive step initiated at time 0 and a negative step of equal amplitude initiated at t1. You might conclude that it is necessary only to analyze the response to each step and then to sum the two at the output to obtain the total response. But such a procedure, when switching is involved, is apt to lead to an erroneous result. The reason is that opening the switch not only produces a negative step but also alters the circuit. As shown in Fig. 3-17 opening the switch removes all possible current paths. Consequently, current, if it is not already zero when the switch is opened, drops instantly to zero. Since no current flows, the capacitor remains charged to the voltage it has when the switch is opened.

In order for the voltages to change, we must provide a path for current to flow. Such a circuit is shown in Fig. 3-18. At time O the current I and capacitor voltage decay and rise as usual. The time constant of the circuit is R1 X C, since R2 does not enter into the charging process. At t1, which we shall assume to be long enough to fully charge the capacitor, the switch is opened. The presence of R2 in the circuit permits current to flow even after the switch is opened. Thus, upon removal of the applied voltage the current, which has been flowing into the capacitor, reverses direction and begins flowing out. At the time the switch is thrown, the capacitor voltage is Vo; thus the maximum value of this discharge current is by Ohm's Law:

$$I\ max\ =\ \frac{Vo}{R1\ +\ R2}$$

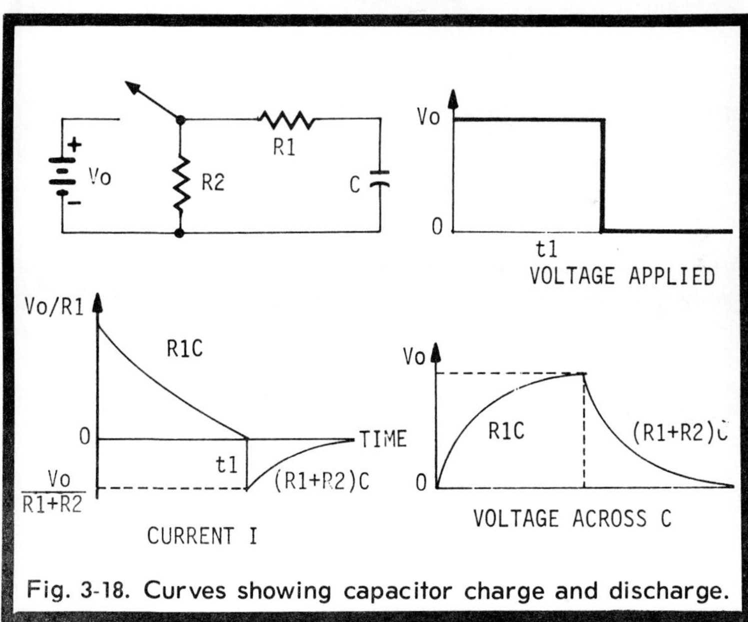

Fig. 3-18. Curves showing capacitor charge and discharge.

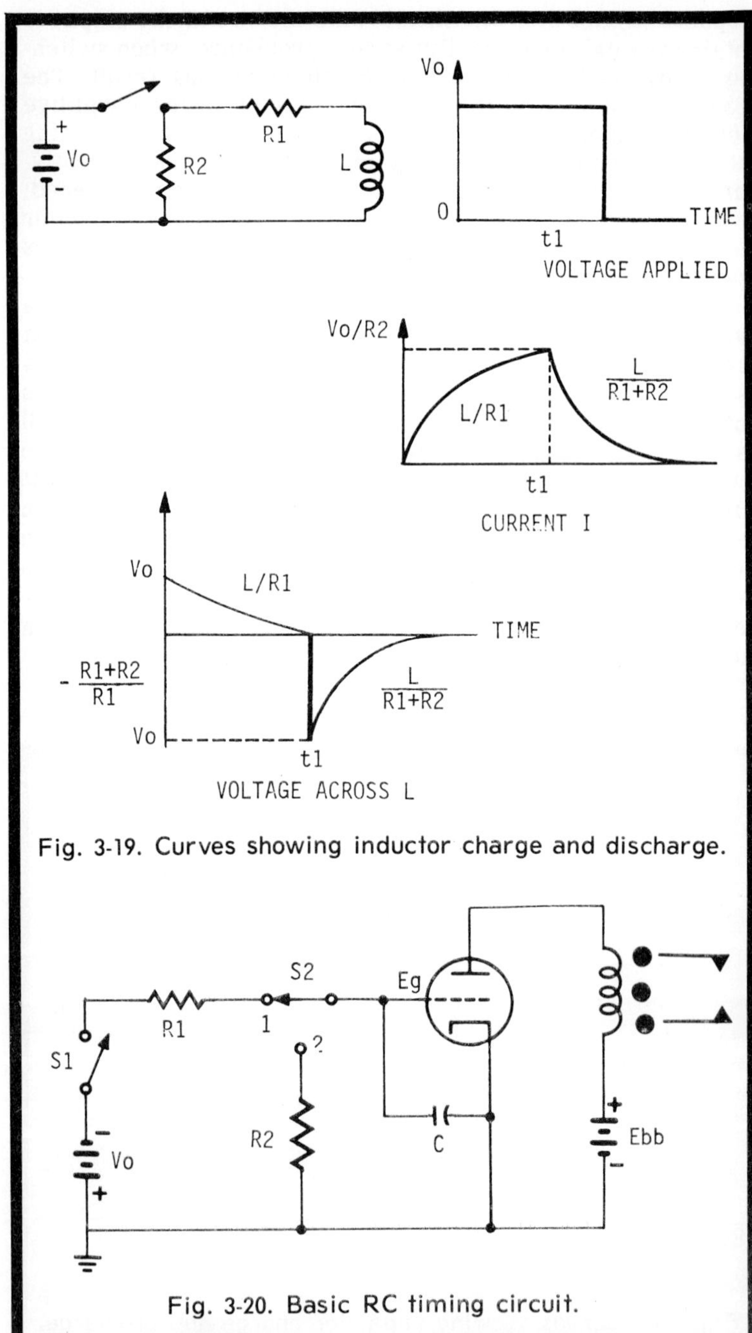

Fig. 3-19. Curves showing inductor charge and discharge.

Fig. 3-20. Basic RC timing circuit.

R1 + R2 is now the total resistance in series with the capacitor: Therefore, it must be used in the calculations. The current and voltage decays during discharge again follow the universal time-constant decay curve, except that the time constant is now $(R1 + R2) \times C$.

We may wonder why, when applied voltage is removed from the circuit, the voltages across the elements do not drop immediately to zero. We must recall that the charging process involved storing energy in the capacitor. When the applied voltage is removed, as in Fig. 3-17, leaving no current path in the circuit, the stored energy simply remains in the capacitor, thus the voltage across it remains constant. When a current path is provided, as in Fig. 3-18, the stored energy is permitted to leave the capacitor. But, energy cannot leave the capacitor instantly, just as it cannot be stored instantly. This accounts for the gradual decay in voltage across the capacitor.

A similar situation exists in a series RL circuit when the switch is closed—current in the circuit gradually rises to a maximum as energy is stored in the magnetic field. Since energy stored in an inductor depends upon current, rather than voltage as in a capacitor, we cannot have the current fall instantly to zero when the applied voltage is removed. Rather, it must decay slowly. When the switch is opened, the inductor itself generates a voltage high enough to arc across the switch contacts, thus preventing the current from dropping instantly to zero. The voltage is generated when the magnetic field begins to collapse around the inductor, and if the switch is opened quickly enough the voltage induced by the decaying magnetic field can be many times the magnitude of the applied voltage, Vo.

In order to prevent arcing across the switch, another current path is provided for the decay in Fig. 3-19. Upon opening of the switch, the current can now decay through R1 and R2. This decay can again be represented by the universal decay curve, except that now the time constant is L divided by $(R1 + R2)$. Furthermore, since current cannot change instantly when the switch is opened, the voltage across the inductor must change sign and have an initial magnitude given by Ohm's Law:

$$V_{ind} = IR_{total} = \frac{V_o}{R1}(R1 + R2)$$

$$= \frac{R1 + R2}{R1} V_o$$

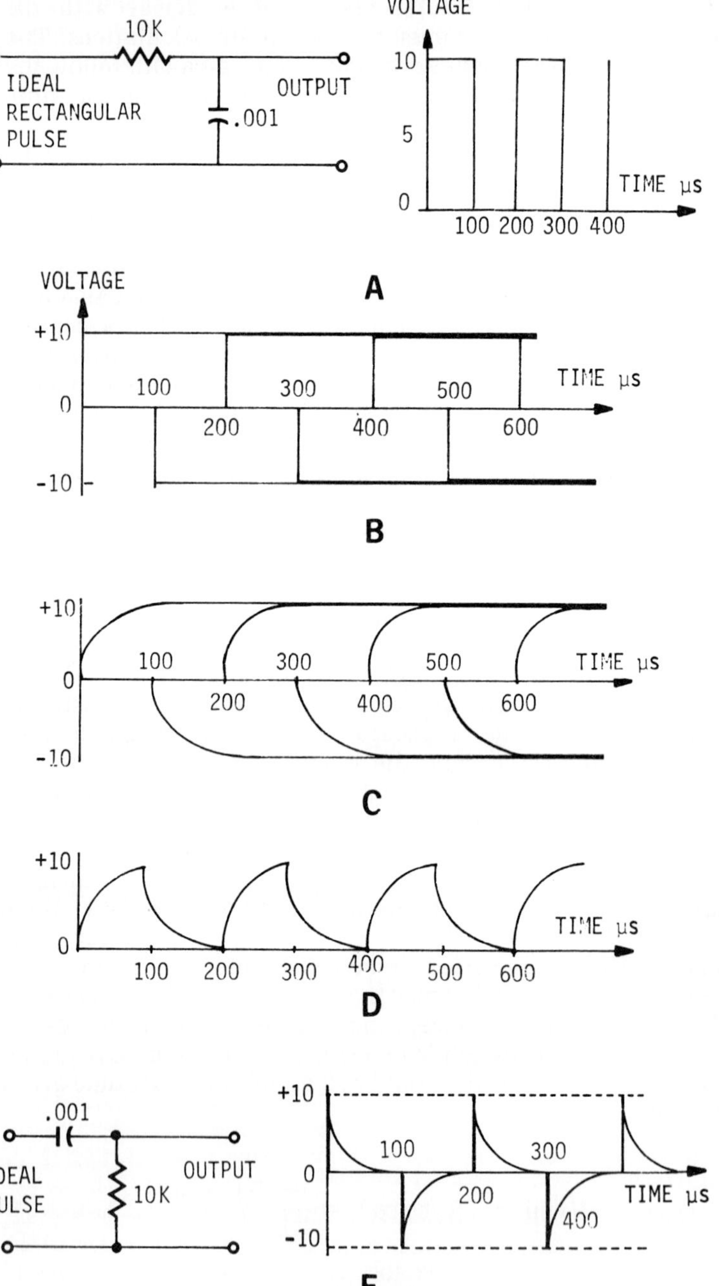

Fig. 3-21. Series RC response curves.

80

If R2 is small enough, the voltage produced by the collapsing magnetic field will not cause arcing across the switch, even though it is larger than the original applied voltage, Vo.

HOW TO DESIGN A TIMING CIRCUIT

Let us now apply our knowledge of RC and LC circuits to the design of a simple timing circuit such as the one in Fig. 3-20. The characteristics of the triode, Ebb, and relay are such that when the grid voltage, Eg, rises to -5v (rises in a negative direction), the relay-coil current drops to the point where the relay contacts open. When the grid voltage falls to -2v, relay current rises sufficiently to again close the relay contacts. The problem is to choose Vo, C, R1, and R2 so that 10 seconds after S1 is closed, the relay opens. Some time later S2 is switched to position 2, and we want the relay to close after 10 seconds more.

Let us begin by making Vo equal 6v. This is not critical, so long as it is greater than the 5v required to actuate the relay. Furthermore, since circuit time constants are R X C, we shall choose 1 mfd for C and set the time constant of the circuit with the resistors. Let's calculate the value of R1. Examining the universal time-constant rise curve in Fig. 3-11, with 6v substituted for 100 percent maximum on the vertical scale, we see that the 5v point, which is required to actuate the relay, occurs at five-sixths of maximum, or 83 1/3 percent and corresponds to a time of 1.79 time constants. Since we want the relay to actuate at 10 seconds, we need 1.79 time constants related to 10 seconds or:

$$1 \text{ time constant} = \frac{10}{1.79} \text{ s} = 5.58 \text{ s},$$

thus we want R1 X C to equal 1 time constant or 5.58 seconds or:

$$R1 = \frac{5.58 \text{ s}}{0.000001 \text{ farad}} = 5,580,000 \text{ ohms},$$

or:

$$5.58 \text{ megohms}$$

In the same manner we can choose R2, working on the assumption that switch 2 is not thrown to position 2 until the grid voltage (Eg) has reached 6v. Then when switch 2 is thrown to position 2, capacitor C, which is charged to 6v, discharges through resistor R2. Thus, we use the universal decay curve and again let its maximum, or 100 percent point be 6v. The relay will close again when Eg reaches 2v, which is 1/3 or 33 1/3 percent of maximum. On the decay curve 33 1/3 percent corresponds to 1.10 time constants. Thus we want:

$$1.10 \text{ time constants} = 10s$$

or:

$$1 \text{ time constant} = \frac{10 \text{ s}}{1.10 \text{ time constants}} = 9.09 \text{ s}$$

$$R2 \times C = 1 \text{ time constant} = 9.09 \text{ s}$$

$$R2 = \frac{9.09 \text{ s}}{0.000001 \text{ farads}} = 9,090,000 \text{ ohms},$$

or:

$$9.09 \text{ megohms}$$

Thus, we have all the component values needed to meet the requirements of the timer.

RL AND RC CIRCUIT PULSE RESPONSE

We now examine the response of an RL or RC circuit to a voltage waveform which can be broken down into the sum of step voltages. Generally, this waveform is the output of a generator or amplifier, so that there is no change in circuitry associated with each step as there is when switches are used to generate step voltages. Such analysis, because no circuit changes occur, requires only a single time constant to use in plotting the response to all steps.

Finding the Output Waveform

As mentioned before, we now want to find the response of the circuit to each of the step components of the input and then to sum the response components to obtain the total response waveform. Let us consider the specific example in Fig. 3-21, and determine the response of a series RC circuit with output voltage across the capacitor when a periodic ideal rectangular pulse input is applied. Although this method of analysis is applicable to all types of waveshapes, the rectangular pulse is used here, since it is one of the most used waveforms in electronic circuitry.

As Fig. 3-21A, indicates, the input consists of ideal pulses (zero rise and decay times) with equal rest and duration times of 100 microseconds each. This corresponds to one pulse every 200 microseconds, or 5000 pulses per second. Furthermore, the time constant of the circuit is:

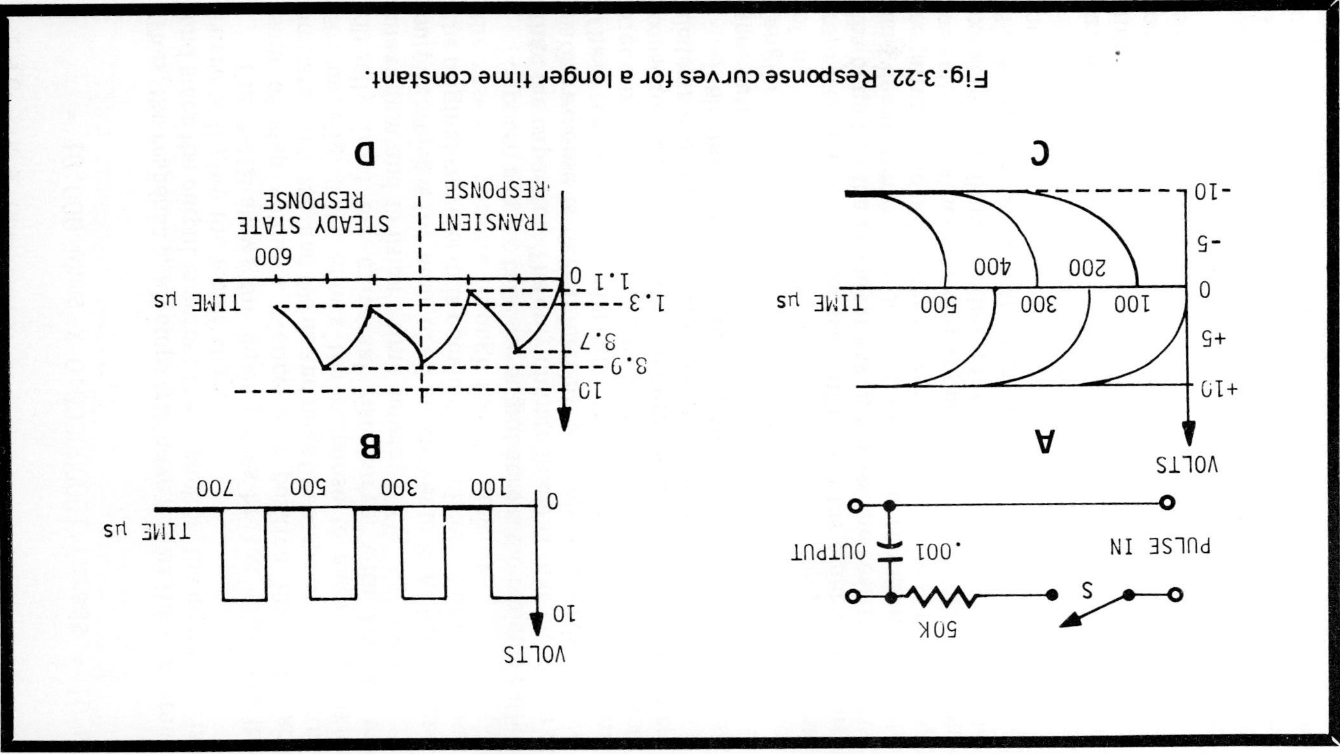

Fig. 3-22. Response curves for a longer time constant.

$$1 \text{ TC} = R \times C$$

$$= 10{,}000 \text{ ohms} \times 0.000000001 \text{ farads} = 10 \text{ us}$$

Thus, the response to all steps will have a 10 us time constant, and since the output is across the capacitor, the universal rise curve will give the exact shape.

Fig. 3-21B shows the input pulses broken down into the sum of step voltages. Notice that positive 10v steps are initiated at 0, 200, and 400 us and negative steps at 100, 300, and 500 us. Fig. 3-21C shows the response to each step taken directly from the universal rise curve, with 10v as the maximum and 10 us as 1 time constant. Fig. 3-21D shows the output, which is the sum of all the curves in B. As you can see, the original waveform is considerably distorted at the output, the result of passing through the RC network.

Suppose now we take the output across the resistor rather than the capacitor. The only change made is that the shape of each response is determined by the universal decay curve. These response characteristics are shown in Fig. 3-21D, and since none overlap, this plot also represents their sum and, consequently, the output voltage. Again notice the severe distortion of the original rectangular waveform.

It should be noted in this example that the distortion when the output is taken across the capacitor is mainly a lengthening of the rise and decay times. This is an indication of poor high-frequency response, and, in fact, the circuit represents a simple low-pass filter. On the other hand, when the output of the circuit is taken across the resistor, the rise and decay times (the decay time being represented by the negative-going pulse) are not zero, as they were in the original waveform. Rather, the output pulse decays rapidly during the constant-amplitude portion of the input pulse, as shown in Fig. 3-21E. This is an indication of poor low-frequency response, and the circuit represents a simple high-pass filter.

In this example notice that the output voltage contains no transient; that is, successive pulses are identical, because the time constant of the circuit is short compared with the input pulse rest time. By "short," it is meant that the circuit time constant is less than one-fifth of the pulse rest time.

Fig. 3-22 shows the effect of lengthening the circuit time constant to 50 us, which is done by increasing the resistance to 50,000 ohms. If time 0 (the moment the switch is closed) is the beginning of a positive pulse, the circuit response to the step voltages will be as in Fig. 3-22C. Summing all the step voltages gives the voltage output as shown in Fig. 3-22D. For one thing, notice that because the pulse duration is not long enough for the capacitor to ever become fully charged to the generator voltage, the steady-state output voltage never reaches an amplitude equal to the generator pulse voltage, nor does it decay to zero during the pulse rest time. The steady-state output voltage works between the peak values of 8.7 and 1.3v. Unless the pulse period time is at least five times the circuit time constant, the output voltage will be of lower amplitude than the generator pulses.

Fig. 3-22D shows that there is a small transient effect. The top peak of the first pulse only reaches 8.77 (against a steady-state value of 8.9v) and the bottom peak at 200 us is 1.1v (against a steady-state value of 1.3v). If the time constant had been longer, the transient effect would have been more pronounced and would have lasted perhaps for many cycles. You should verify the transient and steady-state voltage values in Fig. 3-22 by finding the step voltage values at C at key times from the universal time-constant curve and then summing for the composite voltage value.

To summarize the significance of time constant in an RC circuit with respect to the ability of the circuit to reproduce rectangular pulses, if the RC network is of the low-pass filter

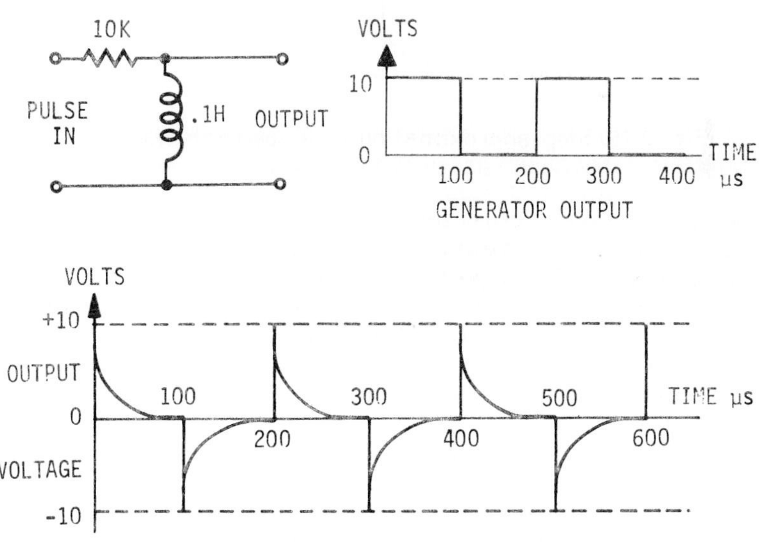

Fig. 3-23. Series RL response curve.

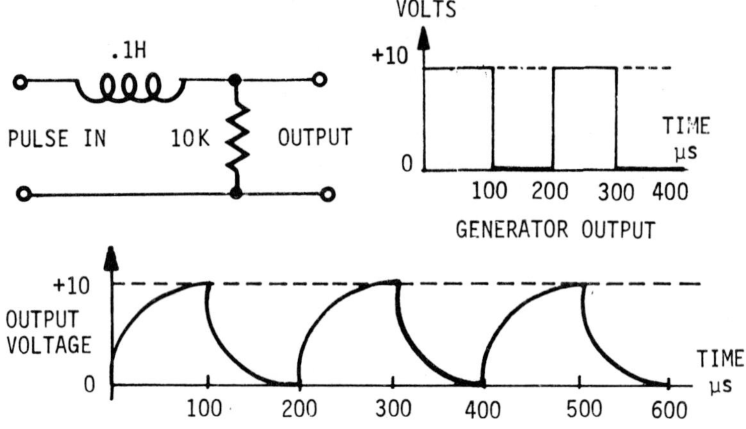

Fig. 3-24. RL circuit response curve.

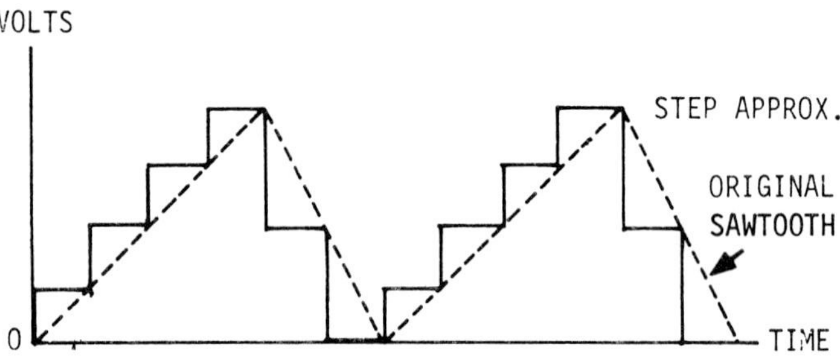

Fig. 3-25. Step approximation of a sawtooth wave.

type (output taken across the capacitor) as in Fig. 3-21, the shorter the time constant, the closer the output voltage waveform will be to a perfect rectangular pulse. If the RC network is the high-pass type (output taken across the resistor) as in Fig. 3-21E, the time constant of the circuit must be many times the pulse period if the circuit is to pass the rectangular pulse of the generator without significant distortion.

The same type of analysis that holds for RC circuits is applicable to RL circuits. Rather than go through the entire procedure of finding the response from component steps, we shall show only the result. Consider first Fig. 3-23, which shows the RL network with the output taken across the inductor. The applied pulse is the same as that which we considered for the RC case. Furthermore, L is 0.1 henries and R is 10,000 ohms, so the time constant is:

$$ITC = \frac{L}{R} = \frac{0.1 \text{ henries}}{10,000 \text{ ohms}} = 0.00001 \text{ s} = 10 \text{ us}$$

This is the same time constant as was used for the RC case. As Fig. 3-23 indicates, the output is exactly the same as it was with the RC circuit, when the output appeared across the resistor.

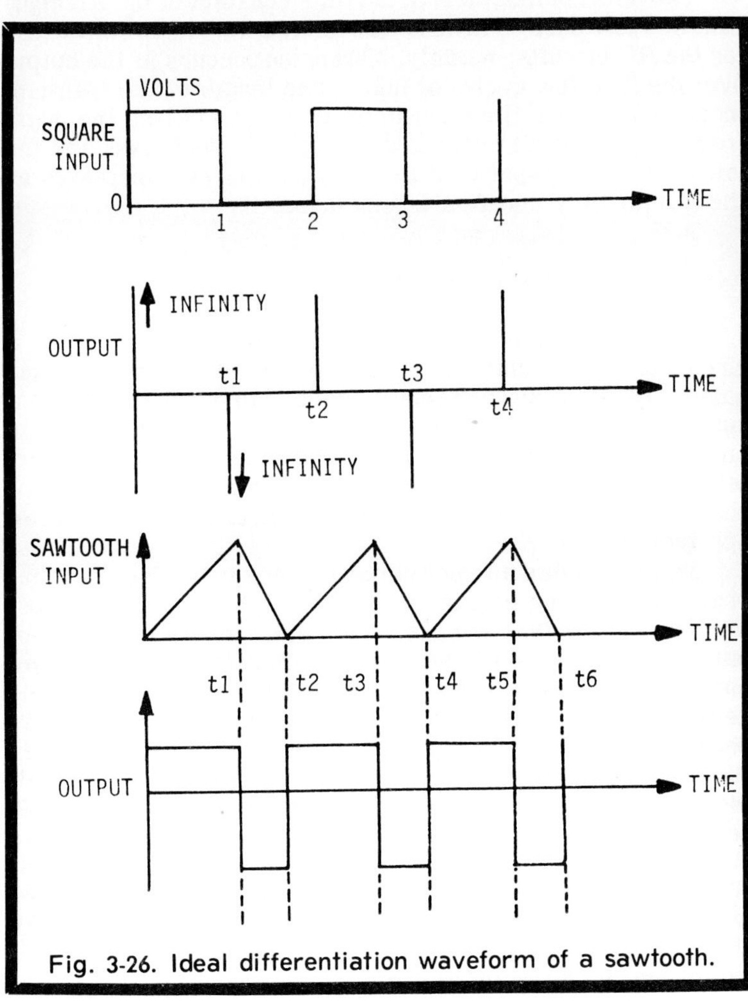

Fig. 3-26. Ideal differentiation waveform of a sawtooth.

If we examine Fig. 3-24, where the output is present across the resistor, we find that the output is identical with that obtained with the RC circuit output across the capacitor. The conclusion we can draw here is that in a series RL circuit the output across the inductor is equivalent in response characteristics to a series RC circuit where the output is taken across the resistor, provided both circuits have the same time constant. Both are simple high-pass filters. Similarly, a series RL circuit that takes the output across the resistor is equivalent in response characteristics to a series RC circuit that takes the output across the capacitor if the circuit time constants are equal. Both in this case are simple low-pass filters.

The effect of lengthening the time constant of the RL high-pass and low-pass circuits is also identical with that discussed for the RC circuits; namely, a transient occurs in the output over the first few cycles of input. The length of the transient increases as the time constant increases. Also, the high-frequency response of the low-pass circuit decreases and the low-frequency response of the high-pass circuit increases as the circuit time constant is increased.

NONRECTANGULAR WAVES

So far we have been concerned with the response of a circuit to rectangular waves and, in particular, rectangular pulses. These pulses are relatively easy to break down into a sum of step voltages. Suppose, however, we would like to analyze the response to a sawtooth or some other nonrectangular wave. With the methods we now have, this is impossible; consequently, we must resort to some sort of an approximation.

Fig. 3-25 shows an approximation for sawtooth pulses that breaks the sloping portion of the wave into a series of steps. The more steps that are used, the better the approximation, but the more difficult the final summation of output components. Not many steps are generally needed to get the general shape of the output waveform. Once the waveform has been approximated by steps, the analysis continues in exactly the same manner as before. As an example, Fig. 3-25 shows the complete analysis of the output voltage waveform that will result when the single sawtooth pulse shown is applied to an RC circuit (the output voltage is taken across the capacitor). We have divided the waveform into five steps, but even fewer would be enough. The response of the RC circuit to each of the step voltages is shown. These are added to give the output waveform, which is as shown.

DIFFERENTIATING AND INTEGRATING CIRCUITS

A differentiating circuit is one whose output signal amplitude is proportional to the rate of change at the input. An integrating circuit is one whose input signal is proportional to the rate of change at the output. Fig. 3-26 shows the result of passing a perfect rectangular pulse and a sawtooth pulse through an ideal differentiating circuit. At times 0, t1, t2, t3 the voltage of the rectangular pulse rises instantly; that is, it has an infinite rate of change. Thus, the output voltage is infinite at these times. At all other times the rate of change is zero, and thus the output voltage is zero.

In reality, rates of change can never be infinite, and the output voltage will never really go to infinity. The sawtooth wave has constant rates of change; that's why the output appears to be rectangular. The leading edge rises slowly in the positive direction, giving rise to a small, positive, constant

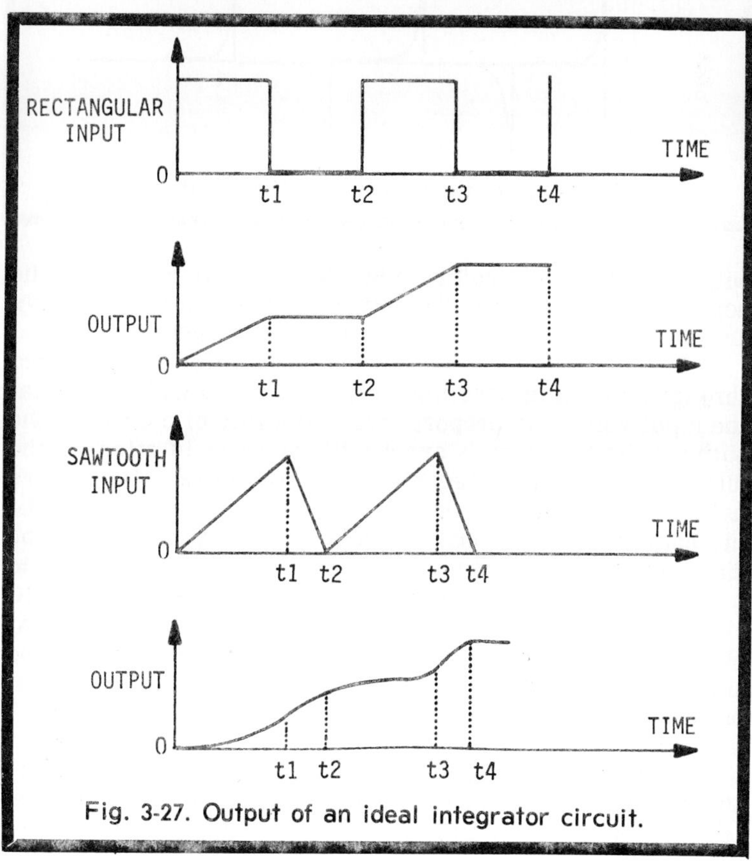

Fig. 3-27. Output of an ideal integrator circuit.

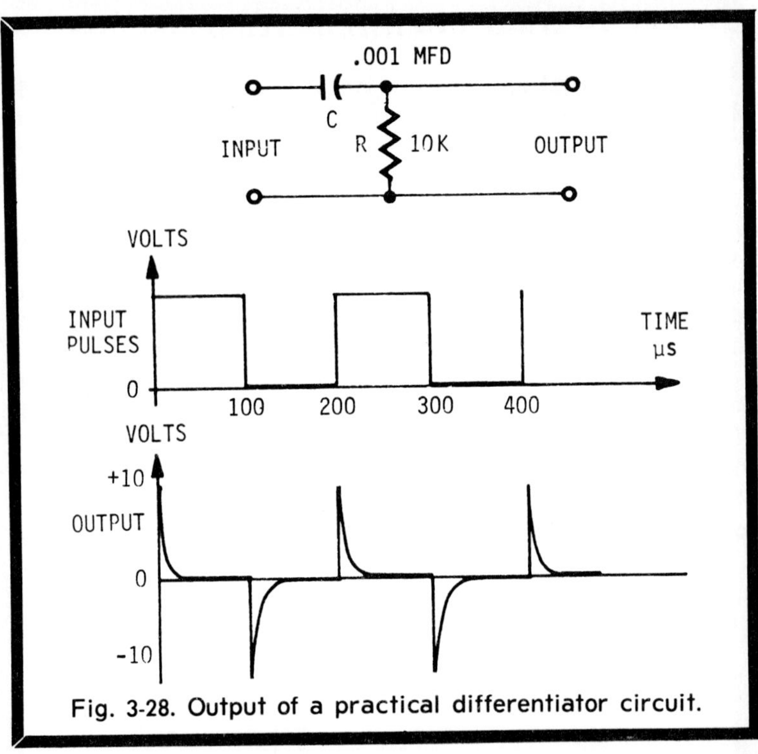

Fig. 3-28. Output of a practical differentiator circuit.

output voltage. The trailing edge decays more rapidly in the negative direction; thus the output voltage becomes negative and greater in magnitude than the positive portion.

Fig. 3-27 shows the same waveforms after being passed through an ideal integrating circuit. Here you will notice that the input voltage is proportional to the rate of change of the output voltage, which is exactly the opposite in effect of the differentiator. True integrators and true differentiators are used mainly in analog computers to perform the mathematical operations of calculus known as integration and differentiation. The circuitry required for true integrators and true differentiators is quite involved. However, simple RC circuits can be designed to have an output approximating that obtained from true differentiators and integrators. These simple RC differentiators and integrators are widely used in pulse work, and they are the ones that are of interest to us. A high-pass circuit, either RC or RL, with an extremely short time constant will differentiate an input signal. By "extremely short" is meant that the time constant is not greater than one-tenth of the shortest time period of the applied waveform.

90

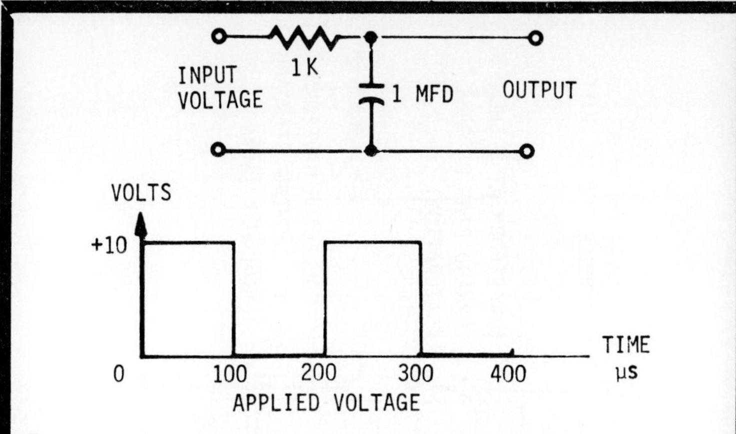

Fig. 3-29. Integrator circuit and applied voltage waveform.

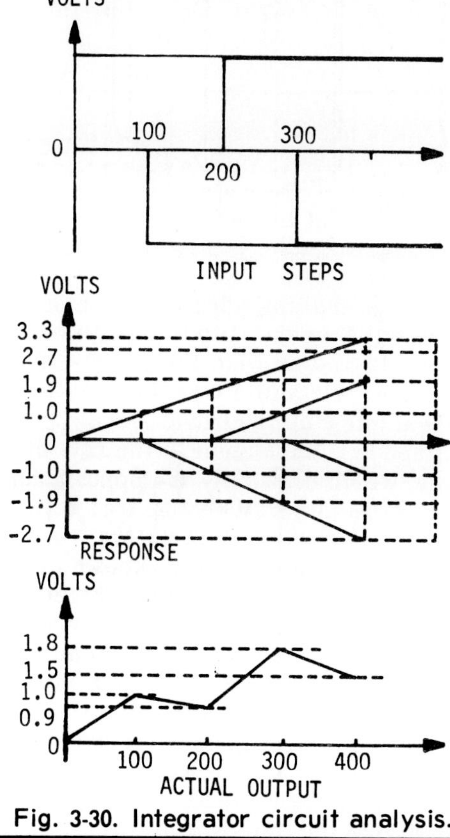

Fig. 3-30. Integrator circuit analysis.

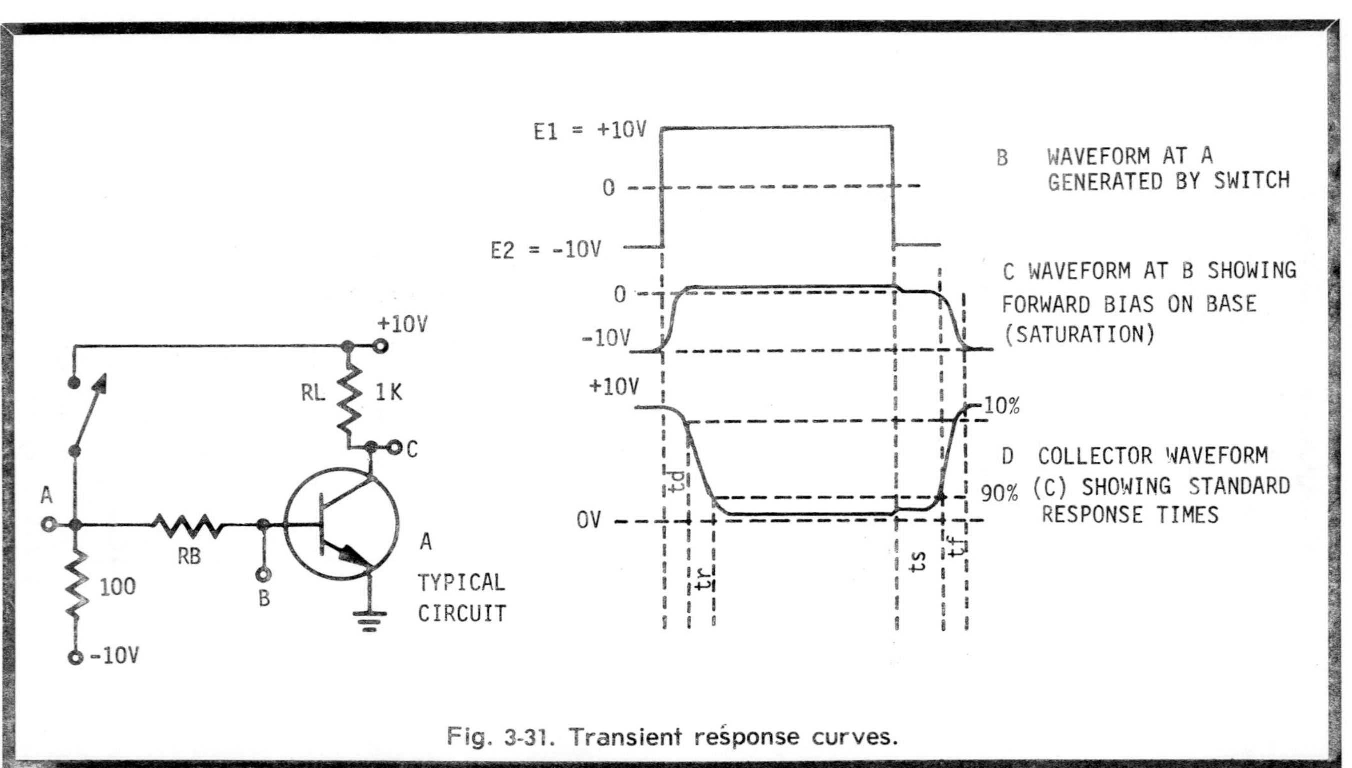

Fig. 3-31. Transient response curves.

Thus, if we wish to differentiate pulses with a 10 us rise time, and if this is shorter than the duration, decay, and rest times, the differentiating circuit should have a time constant not greater than 1 us for reasonably accurate results.

Fig. 3-28 shows the output from a differentiator with a 1 us time constant when a true square wave is applied to the input. The output spikes are quite narrow, but the spike voltage, which would be infinite if from a true differentiator, cannot be greater than the input pulse voltage, which is 10v. To see why the spikes shown are obtained from the circuit of Fig. 3-28, compare with Fig. 3-26 and notice that, as the time constant gets shorter and shorter, the output waveform becomes distorted more and more from the input waveform, until, with a short enough time constant, the output becomes two spikes.

Another way to see why you get spikes out of the differentiator circuit in Fig. 3-28 is to note that there is a voltage out only when current flows through R, because otherwise there would be no voltage drop across R. When the leading edge of a pulse is applied to the input, current rushes through R to charge C. But because of the small time constant, C is charged almost instantly, so that the current flows for but a moment. On the trailing edge of the spike, C discharges and

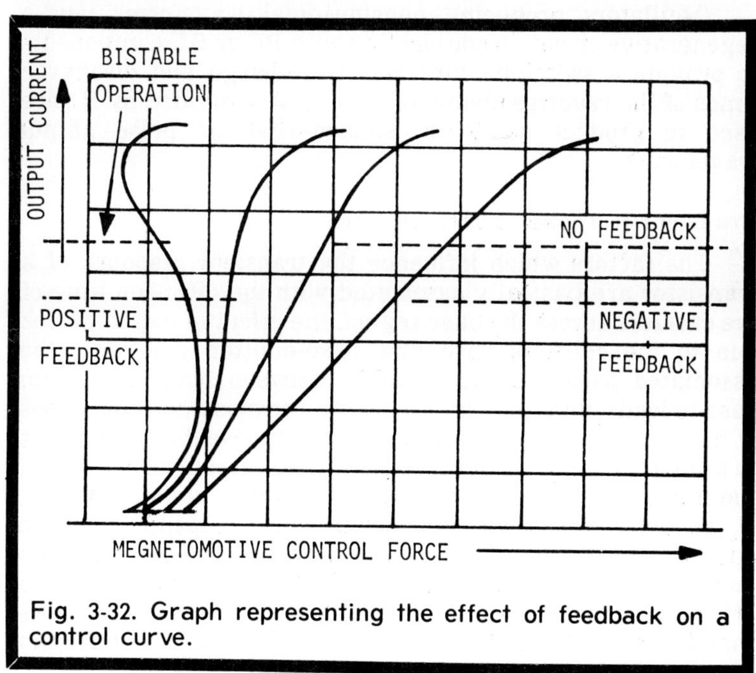

Fig. 3-32. Graph representing the effect of feedback on a control curve.

current flows the opposite way through R to give the negative spike shown. Since C is discharged in a moment, current flow through R is again of very short duration. In contrast to the differentiator, one of the simplest types of integrating circuits is a low-pass circuit, either RL or RC, with an extremely long time constant. By "extremely long," it is meant at least 10 times the longest time period of the applied waveform. Thus, in order to integrate the 100 us pulses that we have been considering with reasonable accuracy, the circuit should have a time constant of at least 1000 us or longer. To show that a low-pass circuit with a long time constant performs as an integrator, consider the RC circuit in Fig. 3-29 with a 1000 us time constant and 100 us pulses as before. Fig. 3-30 shows the input broken down into steps, the response of each step, and the sum to give the total output. As we can see, the output approximates that of the ideal integrator in Fig. 3-27, because with such a long time constant we are always working near the bottom portion of the universal time-constant rise curve. Here, the curve is very nearly a straight line, and, as can be seen from Fig. 3-30, it is this property that makes the circuit operate as an integrator.

NONSINUSOIDAL OSCILLATORS

Oscillators producing nonsinusoidal waveforms use a regenerative circuit in conjunction with RC or RL components to provide a switching function. The charge and discharge times of the reactive elements (R X C or L divided by R) are used to produce sawtooth, square-wave or pulse output waveforms.

Transient Response Characteristics

The factors which influence the transient response of a transistor are basically associated with the diffusion time of the carriers across the base region, the effect of capacitances due to the collector-base and base-emitter junctions, the associated parasitic capacitances between leads and from case to leads, and, just as important, the operation conditions of the circuit. In predicting transient response it is convenient to think of the turn-on delay time, td, the current rise time, tr, the storage or turn-off delay time, ts, and the current fall time, tf, as dependent variables whose values depend upon the operating conditions of the device as well as the capacitances and diffusion parameters. These latter quantities are also affected by the operating conditions of the circuit. It follows that the calculations of the time intervals (td, tr, ts, and tf) can be very complex.

94

Definition of Time Intervals and Currents

The time intervals defined in Fig. 3-31 are almost universally accepted for measurements and, therefore, require no further explanation beyond pointing out that the 10 percent and 90 percent points of the collector waveform are taken as the points at which measurements are to be made. The collector waveform, of course, is the voltage from collector to emitter. For most calculations we shall use collector current rather than collector voltage as the reference, thereby avoiding some difficulty with what is meant by rise time, tr. The voltage is falling during tr but, of course, this is due to the fact that the current is increasing (or rising) during this interval.

Turn-on Delay, Td

Consider the circuit in Fig. 3-31 with the switch in its open position. Under static conditions there exists only a -10v source connected to the base through the 100-ohm resistor in series with Rb. Thus, the base is reverse-biased at -10v (plus a very slight voltage drop due to leakage currents), and the transistor is off. The collector, therefore, must be at +10v (minus a very small voltage due to collector leakage, Io) and the total voltage between collector and base is +20v. Any capacitance between base and collector, therefore, is charged to 20v, and any capacitance between base and emitter must be charged to 10v. Since the transistor is cut off, it is effectively not in the circuit.

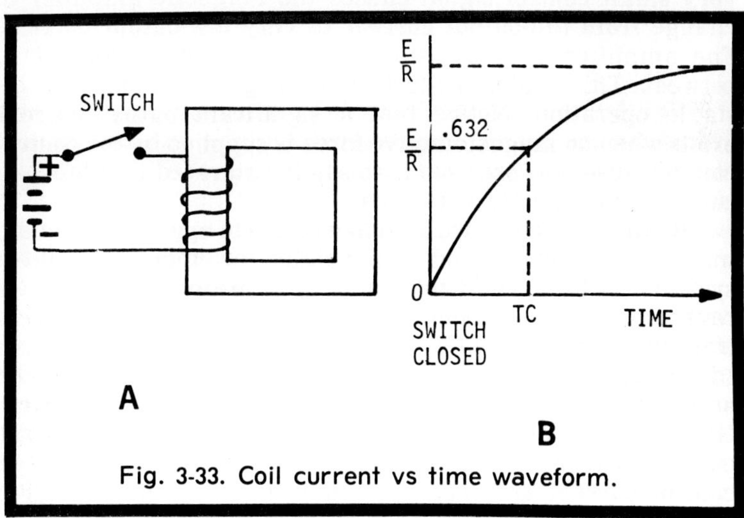

Fig. 3-33. Coil current vs time waveform.

At the instant of switch closure, the voltage at the base cannot change immediately because of the capacitance associated with the base. This means that effectively 20 volts has been placed across Rb thus making Ib equal 1 ma as indicated in Fig. 3-31. Until the base-emitter voltage vanishes, there is no way the transistor can turn on. As current continues to flow, the transistor becomes forward biased, base to emitter, and begins to turn on. This occurs when Vbe approaches about +0.1v for germanium and +0.5v for silicon. Beyond this point the base-emitter diode acts as a clamp so that the voltage cannot continue to rise at the base. Since 0.5v is very small compared to 10v, the final current will be about 0.5 ma as indicated in Fig. 3-31.

MAGNETIC AMPLIFIERS AND REACTORS

As with other types of amplifying devices, positive or negative feedback may be added to magnetic amplifiers. When the feedback winding is connected for regenerative or positive operation, the flux produced by the load current in the winding is in the same direction in the core as the flux produced by the control winding. Thus, the load current aids the control current so that less control current is necessary for a desired change in output. Power gain is increased by the positive feedback, but the undesirable effect is increased response time to a change in the control signal.

If the regenerative feedback in a magnetic amplifier is increased above a certain point, a condition occurs where a very small control signal causes the magnetic amplifier to change from full-output current to very low-output current. The amplifier does not stabilize at any output current in between. This condition, illustrated in Fig. 3-32, is known as bistable operation. Notice that a significant output current exists when no magnetomotive force is supplied by the control coil, because the core used is so easily saturated that the load current alone partly saturates it.

Bistable reactors are used like relays to operate solenoids, magnetic clutches, and annunciator or indicator alarm systems. When the direction of the feedback winding is reversed, the flux produced by the output current subtracts from the flux produced by the control current. This negative (degenerative) feedback thus causes the power gain of the amplifier to be lower than it is without any feedback. However, with negative feedback the waveform of the output current is more nearly sinusoidal and the relation between control current and load current is more linear. Negative feedback also shortens the amplifier's response time.

Time Constant and Response Time

Flux produced by the control coil is proportional to the number of turns (N) of the coil and the current (I) in the coil. For a magnetic amplifier to be controlled by a very small current, the control winding must contain a large number of turns. The inductance (L) of a coil, such as the control winding, is proportional to the square of the number of turns (N^2). As an example, if a coil has four times the turns of another of the same type, it has sixteen times the inductance. Thus the control coil of a magnetic amplifier is highly inductive. This inductance causes a delay in the response of the coil to a changing control signal. Fig. 3-33A shows a control winding to which a battery is connected. Switching on the battery voltage is the same as rapidly changing the control signal. The current in the winding after the switch is closed is shown in Fig. 3-33B. The inductance of the coil keeps the current from jumping immediately to the final value, E divided by R, which would be obtained from Ohm's Law, where R is the resistance of the coil. Instead, current rises to the final value at a rate determined by the inductance and the resistance of the circuit. Time constant and response time are terms used to indicate the rate at which the current rises. Time constant is the time which it takes the current to rise to 63.2 percent of its final value. This point is indicated on the curve of Fig. 3-33B. This time in seconds, in an inductance-resistance circuit, is always equal to the inductance in the circuit by the resistance. Expressed algebraically: Time constant (seconds to reach 63.2 percent of the final value) equals L divided by R.

To find the time constant of the control coil of a magnetic amplifier, make use of the fact that the inductance of a coil is

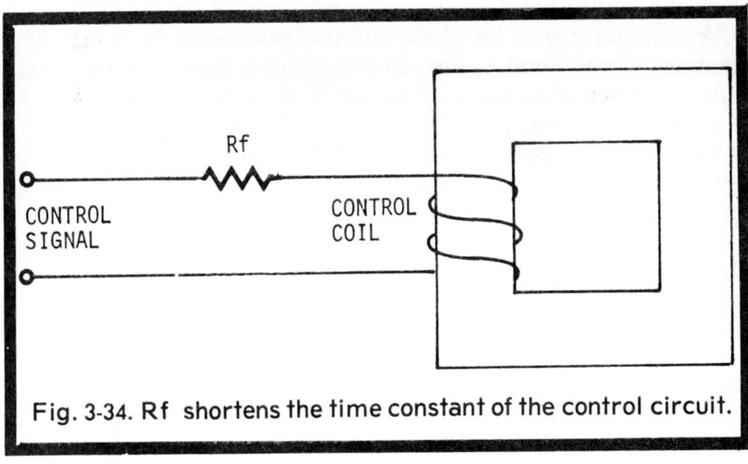

Fig. 3-34. Rf shortens the time constant of the control circuit.

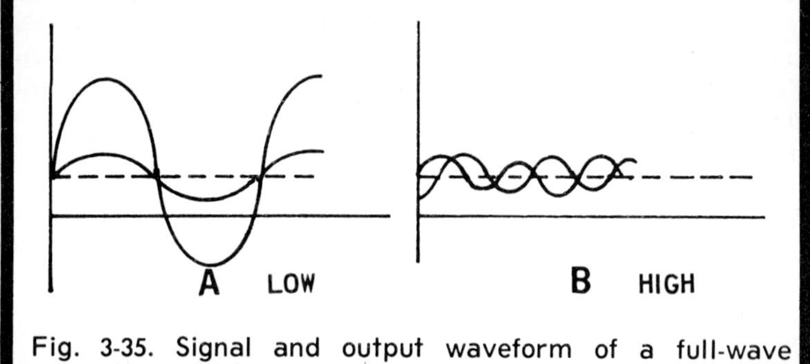

Fig. 3-35. Signal and output waveform of a full-wave magnetic amplifier.

equal to the square of the number of turns multiplied by a contant which depends upon the properties of the amplifier core. If N is the number of turns and C1 is the core constant, the inductance is

$$L = N^2 C1.$$

Substituting this value for L in L divided by R, we obtain the following formula for the time constant for a control coil:

$$\text{Time Constant} = \frac{C1N^2}{R}$$

Response time is the time taken for the current in a coil to reach a given percentage of its final value. This percentage must be given when the response time is specified. For instance, a response time of 0.06 seconds for the current to reach 95 percent of its final value might be given. A response time of 0.02 seconds might be given for the same coil to reach 63.2 percent of the final value. In the second case, the response time is the same as the time constant. Sometimes the response time or time constant is given in cycles of supply frequency instead of in seconds. When this is the case, we would divide the time constant (or response time) in cycles by a frequency in cycles per second to obtain the time in seconds. Thus a response time of 9 Hz at 60 Hz would be 9 divided by 60 equals 0.15 seconds. Conversely, if the time constant is given in seconds, to obtain the time constant in cycles, multiply times in seconds times the frequency in cycles per second.

Since control-coil current is delayed in reaching its final value when the control signal changes, the output current is also delayed the same amount. An additional delay also is caused by core losses, output circuit inductance, and other factors, but this is usually small compared with the delay caused by the control circuit inductance. From TC equals L divided by R, it is seen that the time constant of an amplifier can be shortened by increasing the resistance of the control circuit (R). This can be done by adding an external resistor, called a forcing resistor, in series with the control winding. In Fig. 3-34, Rf is the forcing resistor and Rc is the control coil resistance. Thus:

$$\text{Time Constant} = \frac{L}{Rf + Rc}$$

However, as Rf is increased to shorten the time constant, the control signal voltage must be increased to maintain the same control current. Thus, more power is then consumed by the control circuit for the same power output, and the power gain of the magnetic amplifier is lower.

Frequency Response

Although the control signal of a magnetic amplifier must be DC, it can be varied sinusoidally to produce a sinusoidal variation in the output current. As the frequency of the input

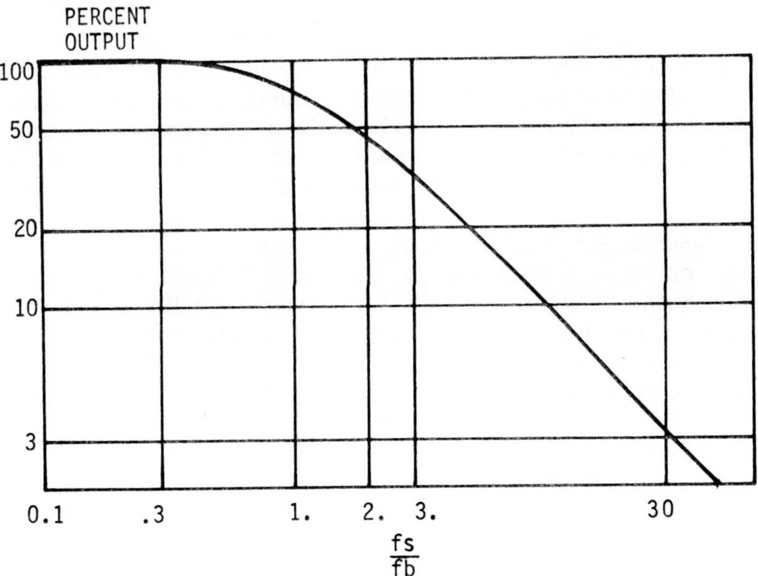

Fig. 3-36. Curve showing the signal-frequency-to-break-frequency ratio.

signal is increased, however, the output current lags behind the signal, and the magnitude of the output current is less. This is illustrated in Fig. 3-35, which pictures the signal and output waveforms of a full-wave magnetic amplifier. In the first case (Fig. 3-35A), the signal frequency is low, and the output current envelope is in phase with the input. In addition, the gain of the amplifier is nearly as high as with a steady DC input.

In case B (Fig. 3-35) the signal frequency is several times higher. The output phase lags the input, and the magnitude of the output is much less. The signal frequency at which the amplifier output voltage lags the signal voltage by 45 degrees is called the break or corner frequency. At this point, output voltage is 70.7 percent of that obtained with a steady DC signal input. This can also be expressed by saying that the output is 3 db down at this point. The break frequency can be found from the time constant by:

$$fb \; = \; \frac{f}{2\pi TC}$$

where fb equals break frequency in Hz
f equals AC power supply frequency, Hz
TC equals time constant in cycles of supply frequency

The break frequency roughly indicates the frequency range of the amplifier. To see what the output of the magnetic amplifier is at some other signal frequency, the curve of Fig. 3-36 is used. The abscissa is the ratio of the signal frequency being amplified, fs, to the break frequency of the amplifier, fb. The ordinate is the percent output of the amplifier compared to the output with a DC input of equal RMS voltage. Notice that both scales are logarithmic for more accurate reading. From this curve it is seen that as the signal frequency increases above the break frequency (i.e., above 1 on the graph abscissa), the output decreases rapidly. A signal frequency of twice the break frequency (2 on the abscissa) will be down to 48 percent output and a signal of ten times the break frequency will be down to 10 percent output.

Magnetic amplifiers show poor frequency response when compared with vacuum tube or transistor amplifiers. Usually they are confined to applications requiring amplification of frequencies below 100 Hz. However, the above equation states that the break frequency is proportional to the frequency of the power supply from which the magnetic amplifier is operated. Thus, the way to increase the frequency response is to increase the supply frequency. Most units that are used in military or in aviation are operated from a 400-Hz power supply. Magnetic amplifiers have been built experimentally to amplify up to 15 kHz, when operated from a 30,000-Hz source.

Construction of Magnetic Amplifiers

Two types of cores, rectangular and toroidal, are used in the construction of magnetic amplifiers. The rectangular cores consist of a number of punched laminations, stacked and riveted together. The toroidal core is made up of a spiral of steel tape. A case is used to hold it in place while the windings are added.

Almost all core materials will distort the magnetic field when the core is subjected to mechanical strain. Much care is used in winding the coils on the core and in supporting the core so that too much strain is not placed on it. Often a box of nylon or phenolic material is employed to support the reactor assembly. In other cases the reactor is placed in a steel box and surrounded with an asphaltic compound.

Chapter 4
Television

Television synchronizing pulses, generated in the studio by a crystal-controlled oscillator operating at a frequency of 31.5 kHz, are locked to the 60-Hz local power line frequency with an automatic frequency control (AFC) circuit. The equalizing pulses are formed by the 31.5-kHz signal, permitting a pulse repetition rate of 31,500 pulses per second. Utilizing a 2-to-1 multi, the original frequency is divided by two to provide the 15,750-Hz pulse for horizontal scanning, and through additional frequency-dividing multis the 60-Hz pulse for comparison to the power line frequency is obtained. The AFC circuit assures correction of the 31.5-kHz oscillator when necessary. The entire system is phase-locked to the line frequency, which prevents ripple in the receiver from becoming noticeable to the viewer.

By the use of several pulse-shaping circuits, the sync pulses are reformed with the steep sides and flat tops required in order to meet F.C.C. standards of good operating practice. The blanking pulses also are formed in the sync generator, along with numerous other timing pulses used by the cameras and studio amplifying equipment. Only 75 percent of the carrier envelope is used for the camera signal and blanking, the remainder being reserved for the sync pulses. Since these are located in the blacker-than-black region, they are never visible on the TV screen.

VERTICAL & HORIZONTAL SYNCHRONIZING PULSES

A vertical sync pulse is transmitted at the termination of each field to guarantee the start of vertical sweep at the camera and receiver together. It is necessary for the receiver to be able to segregate vertical and horizontal sync pulses for routing to their respective scanning generators. The vertical pulse is given a duration time that is nineteen times as long as the horizontal pulse, thus the two can be readily separated with integrating and differentiating circuitry. The rise time and decay times are important factors in evaluating the usefulness of the pulse in a television receiver.

Although vertical and horizontal sync pulses are used for two widely different purposes, they differ only in duration. Both are removed from the square-wave portion of the video signal by the differentiator and integrator. Since the sides of the square wave are vertical, time is not involved. Only during the on time of the pulse are the derived pulses generated, with the polarity reversed during the instantaneous vertical portions. The application of a square-wave voltage to the series RC circuit causes an inrush of current, forming differentiated pulses, and the effect of this current on the voltage developed across the capacitor forms the integrated pulses. The differentiated pulse, therefore, is the instantaneous IR drop across a resistor, due to the capacitor-charging current. The integrated pulse is the voltage developed across the capacitor over a period of time, due to the cumulative current flowing in said capacitor.

BLANKING SIGNAL

Blanking pulses, needed to remove retrace lines, are transmitted at the end of each scanned line and at the end of each field. These pulses swing the grid of the CRT negative to the black level, cutting the beam off until retrace is completed. Sixteen percent of the horizontal line scanning time is

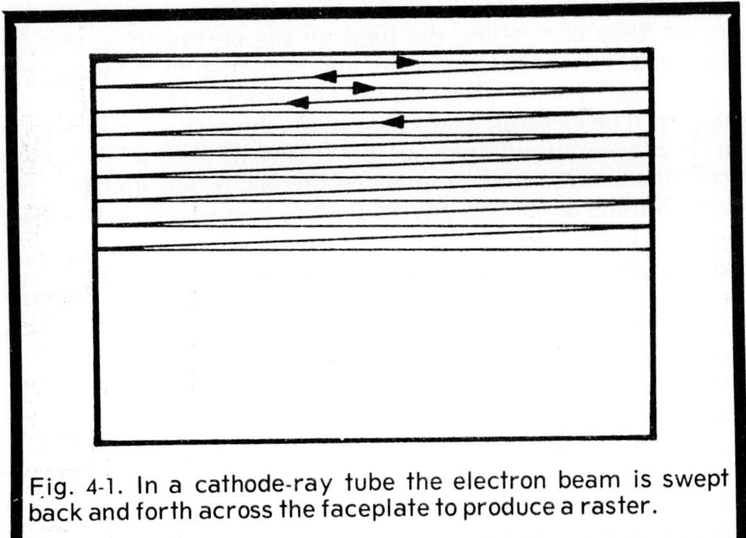

Fig. 4-1. In a cathode-ray tube the electron beam is swept back and forth across the faceplate to produce a raster.

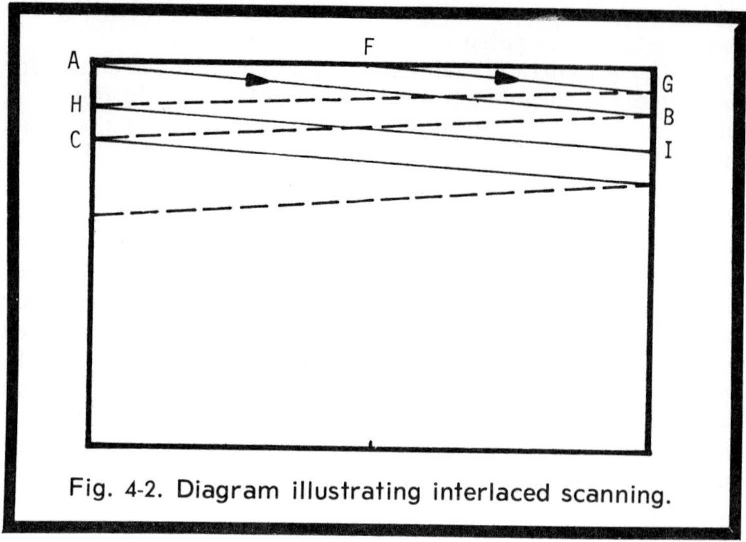

Fig. 4-2. Diagram illustrating interlaced scanning.

used by the horizontal blanking pulse, and eight percent of the frame scanning period is used by the vertical blanking pulse. Thus, 42 horizontal lines are blanked out, leaving 483 active.

SCANNING

Reproducing a transmitted picture in a television receiver requires scanning a cathode ray tube (CRT) with an electron beam while the intensity of the beam is varied by the video signal, thus controlling the light on the screen. U.S. system standards for the transmission and reception of monochrome video make use of horizontal linear scanning in an interlaced pattern. The electron beam first scans the entire viewing area in an alternate (odd-line) line procedure as shown in Fig. 4-1, and then returns to the top of the picture to rescan the same area, but the beam now covers the lines (even) between those covered in the first field. Thus, the first field consists of all the odd-numbered scanning lines, while the second field of each frame covers the even-numbered lines. The two interlaced fields form a frame which is a complete scanning cycle. Field repetition rate is 60 per second, with vertical scanning rate of 60 Hz. The scanning cycle is repeated upon the completion of each frame.

Fig. 4-1 illustrates lines formed by the zig-zagging electron beam sweeping across the screen with uniform velocity, covering all picture elements in one horizontal line, and slowly moving downward with each sweep. The horizontal lines slope downward in the direction of scan as vertical deflection

104

produces a simultaneous vertical scanning motion. Only the lines formed when scanning left-to-right, as seen from the front, appear on the screen. The retrace lines formed as the beam moves from right-to-left, are blanked out by a pulse which applies a negative voltage to the CRT grid, cutting it off. As beam intensity is reduced to zero during retrace periods, the lines do not appear on the screen.

Interlaced scanning, illustrated in Fig. 4-2, starts at Point A in the upper left corner where the scan of the first field beam begins. The horizontal deflection yoke windings pull the beam from Point A to Point B, exciting the screen phosphors in its path to the intensity dictated by the picture signal, forming line AB. The slanting line is due to the steady downward pull of the vertical deflection system at the very slow rate of 16,667 microseconds per field as compared to the horizontal rate of 63.5 microseconds per line. At Point B, the beam quickly returns (cut off by the blanking pulse) to the left side of the screen at Point C. The retrace speed (11.43 microseconds) is much faster than left-to-right scanning; therefore, the slope is very slight. The beam continues slowly downward as succeeding horizontal lines are scanned, until the bottom of the picture is reached. Since all of field one has been scanned, a longer blanking pulse of 833 microseconds (vertical blanking) cuts off the beam while the vertical deflection system returns it to the top of the picture at Point F. Vertical flyback time is very fast compared to the trace, but slow when compared to the horizontal scanning; therefore, as horizontal sweep continues, there is a series of zig-zag lines from bottom to top.

There are 262½ horizontal scanning lines in each field, the last line of field one and the first line of field two are only half-lines, accounting for the fraction in the total. Because of the vertical blanking pulse at the end of each field, several of the last lines are blanked out as the beam is returning from the bottom to the top of the picture. Starting the second field at Point F, half-line FG is first scanned, followed by HI, and the sweep continues until the field has been completed. Thus, the complete cycle consists of two fields, one frame, and 525 lines. About 10 percent of the total lines are lost during blanking and are visible in the raster as a black bar only when the vertical oscillator is out of sync.

The 60 fields-per-second rate is vital because of its relationship to the power line frequency of 60 Hz. Thus any small amount of ripple, as a result of less than perfect filtering of the DC supply, remains stationary in the raster and is not

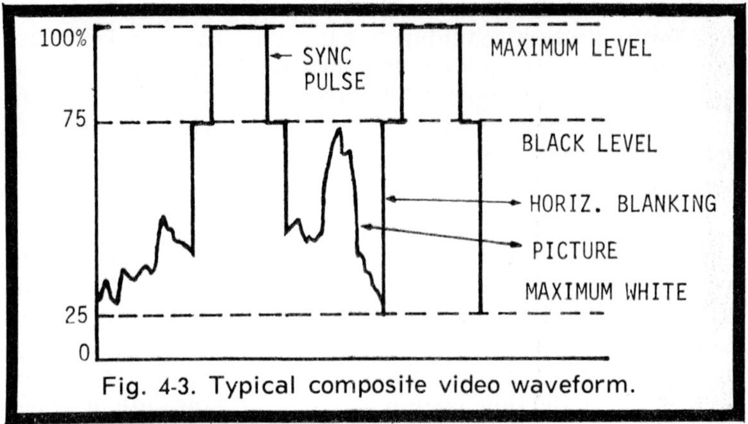

Fig. 4-3. Typical composite video waveform.

noticeable, otherwise the continuous motion of the ripple across the screen would be quite annoying. As an additional plus, the vertical sync is far more dependable when the field frequency is tied to the line frequency. Any residual ripple has a tendency to trigger the vertical sweep and assist in keeping it in perfect synchronization.

Interlace scanning eliminates flicker because the viewing area is covered 60 times (field frequency) per second, and even though the frame rate is only 30 per second, the eye fails to notice the difference at the speed involved. The ratio of the width to the height known as the aspect ratio of the picture is 4 to 3, the same ratio used in motion pictures until the wide screen came along. However, this aspect ratio makes televising of standard films possible without waste, and has been proven to be eye-appealing as well.

COMPOSITE SIGNAL

The timing pulses in the composite video signal guarantee the simultaneous start of the scanning beams in the receiver CRT and the camera tube, as mentioned earlier, and the blanking pulses extinguish the beam during retrace. The picture is brightest when whites are transmitted and the video-signal amplitude is at its lowest point of 12.5 percent. Grays require a progressively higher amplitude until the black level is reached at 75 percent of maximum; the 25 percent above the black level is reserved for sync pulses.

The degree of modulation in the transmitted signal is exactly the reverse of the action at the pickup tube, which has maximum output on whites and none whatsoever on blacks. This reversal is known as negative transmission, and does greatly reduce interference on the receiver screen under less than favorable conditions. Fig. 4-3 shows the detected video

106

signal amplitude and limitations. Applied to the CRT grid, the video signal varies the intensity of the electron beam according to the picture information. As the camera becomes inactive after each horizontal line, the blanking pulse maintains the beam at the black level as retrace is initiated.

Sync pulses are separated from the video signal by a sync separator, which is biased so that an output signal is produced only on peak positive swings of the input through the upper 25 percent level where sync pulses appear. When a transistor is utilized for the separation stage, the function is similar but amplification of the pulses results. After the sync pulses are separated from the video, hroizontal and vertical pulses are filtered by RC circuits; the horizontal is selected by a differentiating circuit and the vertical by an integrating circuit.

HORIZONTAL SYNC DIFFERENTIATOR

Horizontal sync signals are isolated in a differentiating circuit with a short time constant compared to the horizontal pulse width. Application of the total sync pulse to the differentiating circuit charges the capacitor completely very soon after the leading edge of each pulse. The capacitor retains that charge for a period nearly equal to the complete pulse width. As applied voltage is removed, at the trailing edge of each pulse, the capacitor is completely discharged within an extremely short time and a positive peak voltage is available for each leading edge. The trailing edge of the pulse produces the negative peak as a result of the discharge current. Serrations in the vertical pulse provide the necessary differentiated output to sync the horizontal scanning generator during vertical synchronization which is so much slower. As the system is sensitive only to positive pulses at the right horizontal timing, negative sync pulses, alternate differentiated positive pulses produced by equalizing pulses, and the vertical information have no effect on timing of the horizontal generator. As a result, only horizontal sync information appears at the output of the differentiating circuit.

VERTICAL SYNC INTEGRATOR

Vertical sync pulses are isolated in an integrating circuit which has a long time constant in comparison to the 5-microsecond horizontal pulses but short compared to the 190-microsecond vertical pulse duration. No voltage is applied to the RC circuit during the period between horizontal pulses, which is so much longer than the horizontal pulse width itself that the capacitor is able to discharge nearly to zero. When the

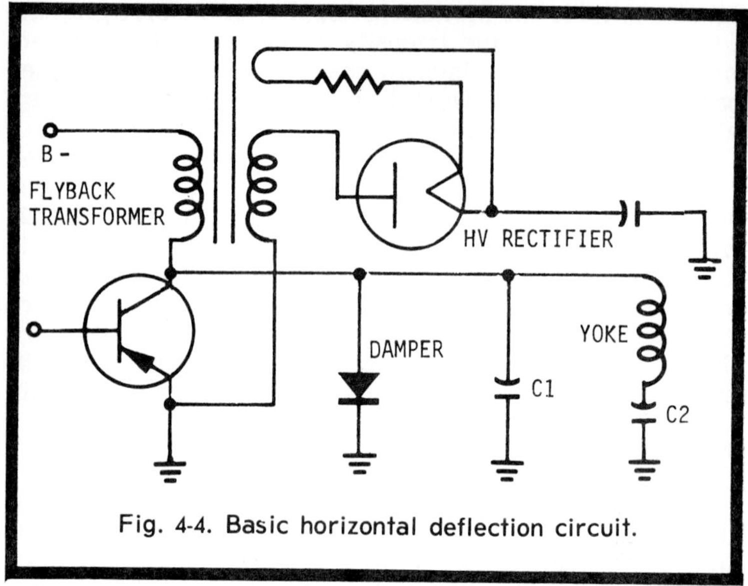

Fig. 4-4. Basic horizontal deflection circuit.

vertical pulse is applied, integrated voltage across the capacitor builds up to that value needed to trigger the vertical generator. The capacitor has an integrated voltage buildup that reaches its maximum amplitude at the end of the vertical pulse, and then discharges to almost zero, producing a triangular pulse. Even though total sync is applied to the input of the integrator, only vertical sync information is available at the output. Equalizing pulses minimize any significant difference in the trailing edge of the vertical pulse between odd and even fields.

HORIZONTAL DEFLECTION

A current that varies linearly with time and which has sufficient amplitude (peak-to-peak) must flow through the horizontal deflection yoke winding to develop an adequate magnetic field to deflect the CRT electron beam. Following the excursion of the electron beam across the face of the cathode-ray tube, it must be extinguished and quickly returned to its point of origin.

The use of a transistor as a switch illustrates the typical horizontal deflection circuit and is shown in Fig. 4-4. Since the transistor cannot act as a switch for the reverse collector current, a damper diode is used to handle this load. A high voltage is generated with flyback (step-up) transformer T1 in parallel with the yoke; leakage inductance, distributed capacitance, and output stray capacitance assist the yoke inductance and retrace capacitance in such a way that the peak voltage across the primary of T1 is reduced and peak voltage across the secondary is increased over expectations as compared to a perfect transformer. Commonly known as third harmonic tuning, a voltage of 1.7 times that expected in a perfect transformer is produced.

Linearity correction for wide-angle picture tubes requires the sweep rate to be slowed at the beginning and end of each scanning line. A suitable capacitor is used in series with the yoke, as in Fig. 4-4, so the DC needed to supply circuit loss is fed through the T1 primary. Thus S-shaping capacitor C2 provides a parabolic waveform, so the trace voltage across the yoke is lower at the beginning and end of the sweep than in the middle.

VERTICAL DEFLECTION

The vertical deflection circuit somewhat resembles a Class A audio stage with a primary need for controlled linearity but burdened with an intricate load and strict low-frequency demands. There are several ways of obtaining such requirements, first by designing an amplifier with a flat low-frequency response all the way down to 1 Hz. But to achieve such a low-frequency response there are at least two major problems—very large capacitors and a large vertical output transformer; thus, that alternative is ruled out. The amplifier, as another alternative, could provide reasonably good response at low frequencies and predistort the overall signal, or by using high enough gain to permit degenerative feedback which in turn would assure linearity. Loop feedback would provide predistortion to eliminate low-frequency problems and nonlinearity would be of little significance with around 25 db feedback. The coupling of undesirable signals in the amplifier would be comfortably reduced or eliminated.

COLOR TV

The N.T.S.C. color TV system which is compatible with B & W transmission, is the standard used in the United States. Color telecasts can be received satisfactorily on any black-and-white receiver and monochrome on any color set, because a B & W receiver ignores the color subcarrier and chroma circuits in a color receiver are cut off during B & W reception.

Chroma IF

The chroma IF amplifier employs but one tube or transistor and is about as simple as such a stage can possibly be,

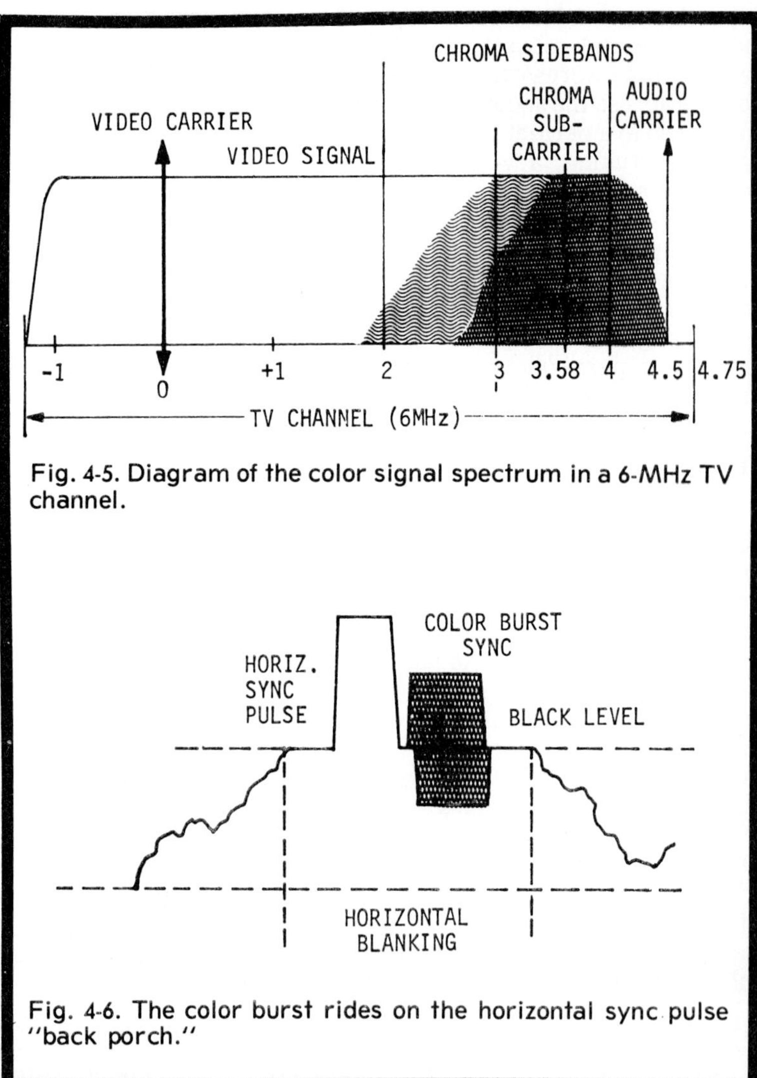

Fig. 4-5. Diagram of the color signal spectrum in a 6-MHz TV channel.

Fig. 4-6. The color burst rides on the horizontal sync pulse "back porch."

except for two differences: The most important is the color-killer voltage inserted at the bypassed end of the chroma take-off coil. The second is the horizontal blanking pulse introduced into the stage via the cathode resistor shared in some receivers by the chroma IF and blanker. When a program is in color, the killer DC injection voltage normally is zero at the chroma amplifier grid. During black-and-white programs, the applied voltage is sufficient to cut off the amplifier.

110

Burst Amplifier

The burst amplifier (keyer) operates only during horizontal retrace time when no 3.58-MHz color subcarrier signals (except the burst) should be at the grid. The "burst" is the color sync, which is amplified, while the other signals, such as chroma or the upper harmonics of the video, are rejected. To accomplish this selective or keyed amplification, the tube (at rest) is biased to cut off with a very large cathode resistor, then driven into conduction with a positive-going pulse at the grid. During the pulse time, the tube operates with just a few volts of bias (thus achieving plate current and gain), but it is cut off completely when the pulse is gone by the voltage stored in the cathode bypass capacitor.

The video carrier is 1.25 MHz above the lower edge of the 6-MHz television channel and, although, video frequencies to 4 MHz are carried, the lower sideband is cut off at 1.25 MHz to conserve space. This system is called vestigal sideband transmission; that is, one of the sidebands is suppressed at the transmitter. The video is amplitude modulated, while the sound portion of the transmission is frequency modulated. At 0.25 MHz below the upper limit of the channel, the audio center frequency (sound carrier) is located. Its frequency swing is restricted to 25 kHz either side of center.

Fig. 4-5 shows the transmitted color signal, the chrominance signals (I and Q) are carried in the open spaces in the Y signal in order to avoid increasing the bandwidth requirements. A 3.58-MHz oscillator (chroma subcarrier) is modulated by I and Q signals, and since balanced modulators are used, the carrier frequency is completely suppressed by the balanced circuit. For color transmission, a color-burst signal (sync) is needed and is impressed upon the back porch of the horizontal synchronizing pulse with a burst covering 8 to 11 cycles of the 3.58-MHz oscillator signal as illustrated in Fig. 4-6.

Special components are necessary for color, and four signals are taken from the video amplifier. The sync signal is very nearly the same as used in monochrome, the luminous signal (Y), the chrominance signal (I and Q combined), and the color sync mentioned in the preceding paragraph. The color killer circuit produces cut off bias for the chroma amplifier, as described previously, thus eliminating the color dots produced by noise in the color amplifiers. By simply pressing a button in the studio, the color-burst signal is eliminated from the horizontal sync pulse, and its absence triggers the color-killer circuit which in turn biases the color amplifier to cut off.

Chapter 5

Digital Computers

The digital computer utilizes a process of calculation employing discrete digits for performing mathematical operations. Digital devices such as desk calculating machines, telephone dialing systems, and digital-readout frequency counters, operate with distinct or separate units suitable for counting these digits. Numerical values involved are represented by the "state" of flip-flops and other switching configurations within the computer. Calculations are performed by the computer in accordance with the rules of arithmetic pertaining to addition, subtraction, multiplication, and division. Accuracy of the highest level, as would be mandatory in accounting, and information correct to one part per million is readily achieved by these digital devices. The ordinary slide rule, basically an analog or proportional device, would require a length of nearly 900 feet to display equivalent readout accuracy. Fig. 5-1 shows a large scale digital computer system used in data communications applications.

Tiny transistors with minimum power needs and inherent reliability are used extensively in today's digital equipment. As semiconductor technology continues to provide improved capabilities and characteristics, digital circuitry becomes more efficient, faster, and ever smaller. Integrated circuits which take full advantage of microminiaturization processes are leading the way to further advancement in the field, with many horizons still to appear. Since basic circuits and design procedures are covered here, the design engineer or technician will find new answers to previously insurmountable problems.

BINARY NUMBERS

Binary refers to the number system based on a radix of two, a system using only two digits, 0 and 1. Much of the circuitry for processing binary signals is directed to the switch-

Fig. 5-1. Burroughs B6500, a large-scale language processor system designed for the multi-processing, time-sharing environment. (Courtesy Burroughs Corp.)

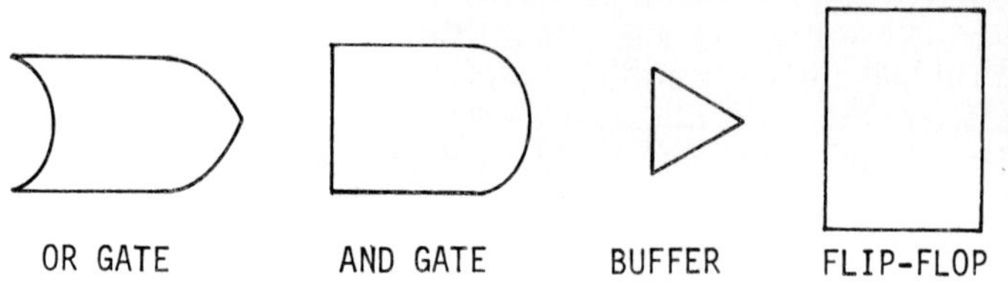

Fig. 5-2. Symbols used in logic circuit diagrams.

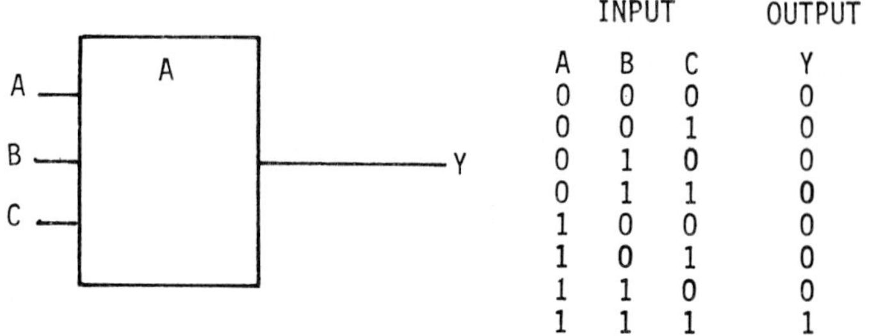

INPUT			OUTPUT
A	B	C	Y
0	0	0	0
0	0	1	0
0	1	0	0
0	1	1	0
1	0	0	0
1	0	1	0
1	1	0	0
1	1	1	1

Fig. 5-3. Three-input AND gate with truth table.

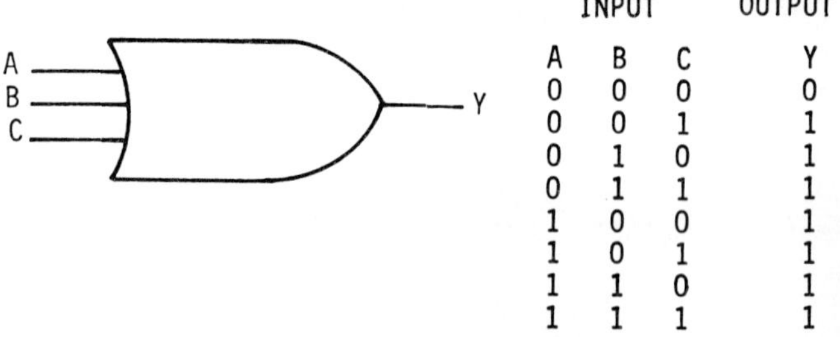

INPUT			OUTPUT
A	B	C	Y
0	0	0	0
0	0	1	1
0	1	0	1
0	1	1	1
1	0	0	1
1	0	1	1
1	1	0	1
1	1	1	1

Fig. 5-4. Three-input OR gate with truth table.

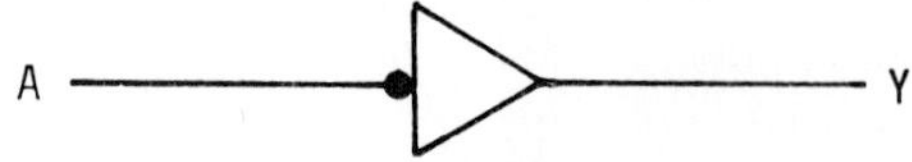

Fig. 5-5. NOT gate or inverter.

ing of the signal from zero to 1, or 1 to zero, and is known as logic circuitry. AND and OR gates are the more important types of logic. Standardized (Government) forms are shown in Fig. 5-2 along with the buffer and flip-flop.

Inputs may number 2, 3, 4, or more, with basic gate functions such as:

AND – All inputs 1, output 1.
NAND–All inputs 1, output 0.
OR–Any input 1, output 1.
NOR–Any input 1, output 0.

The AND gate has more than one input with a single output; the same holds true for the OR unit. But from here on, the two gates no longer look alike as the AND has an output only when all inputs are entered simultaneously; the OR when any input exists. A small o at the output terminal is normally used to indicate inversion has taken place.

Binary indicates two parts, and most computers, data-processing, and automatic control devices make use of the two possible states signified by the numbers zero and one. The inactive state usually refers to the 0 state, and state 1 indicates the active, energized, or stable state. A closed switch is state 1, and an open switch is state 0. The AND gate output is in state 1 if a pulse is present, or in state 0 if none is present. Adding and subtracting with logic circuitry is quite simple when utilizing this two-digit system, i.e., 0 and 1. As we work into actual circuits using transistors, it becomes necessary to alter our original signal state system of voltage or no voltage. Instead, we select two different voltages that are suitable for the particular solid-state units in use.

The terms "positive" and "negative" logic merely indicate whether state 1 is more positive, or more negative than state 0. The binary figures 0 and 1 facilitate tracing and checking out the equipment. Other terms, such as true or false, are frequently used in place of 0 and 1. However, in order to avoid possible mix-ups, remember that true means state 1 and false state 0. This never changes regardless of what the actual voltage or lack of voltage may be.

Moving forward with this new concept, definitions for AND and OR gates must be changed somewhat to conform:

AND–a unit or circuit with an output in state 1 **only** when all inputs are in state 1. If any input signal is in state 0, the output is always in state 0.

OR–a unit with the output in state 1 when **any** input is in state 1. The output is in state 0 **only** if all inputs are state 0.

TRUTH TABLES

Truth tables are standard tools for studying and designing logic circuitry. In order to make certain "truths" more apparent, truth tables for the AND circuit and the OR circuit are shown beside the applicable drawings in Figs. 5-3 and 5-4, respectively.

Truth table input combinations always total one less digit than the number of digits in the binary; for example, 8 combinations for 3 inputs (binary 8 equals 1000), 64 combinations for 6 inputs (binary 64 equals 1,000,000), etc.

NOT GATE

The NOT gate or inverter has only one input as well as one output, and the output is always the reverse of the input. When the input is in state 1, the output is in state 0. Fig. 5-5 displays the symbol for the NOT unit. Notice the small o before the symbol.

BOOLEAN EXPRESSIONS

Boolean expressions simplify the relationship of inputs to outputs in logic circuitry. For example, showing the output of an AND circuit with inputs A and B, we simply indicate the output as A x B (A times B). The OR circuit with A and B inputs, would show A + B (A plus B) for the output. Following the practice as in algebra, we may drop the x sign and signify A x B as AB. The + sign can never be dropped in Boolean expressions, although every other simplifying or shortening procedure is not only acceptable but desirable.

A + B + CD is in state 1 if either A, B, or CD is in state 1; otherwise, the output A + B + CD would be 0. In evaluating the expressions, we simply treat plus and times signs as in regular arithmetic. For example, AB + C equals Y with A in state 0, B in state 1, and C in state 1; Y equals (0 x 1) + 1 equals 0 + 1 equals 1, so the output Y is in state 1. Whenever zero is involved in multiplication, the answer is always zero.

If the input to an inverter is A, we designate the output as A (NOT A); if the input A is state 1, then the output A is state 0; that is, NOT the state of the input. If A is 0, then A is state 1. Actually, most complex logic systems can be built from AND, OR, and NOT gates alone. However, other useful switching units are NAND, NOR, and EXCLUSIVE OR. The latter has two inputs, but only produces a state 1 output when one and not both inputs is in state 1. At all other times, the output is state 0. Fig. 5-6 shows the NAND, NOR, and EXCLUSIVE OR symbols.

116

DIODE SWITCHING UNITS

Relay switching units, quite useful before the discovery of semiconductors, were bulky, slow, and power consuming. Vacuum tubes were also used in systems of earlier design, but we are all well aware of their shortcomings compared to semiconductors. AND and OR gates using diodes need few components; just a diode for each input, a resistor, and a source of power. A complete diode AND circuit is shown in Fig. 5-7. The power source Vs (+50v) is always higher than either of the two signal voltage levels used (-10v and +5v). Power supply voltage Vs is positive for an AND circuit in a positive logic system.

For an OR circuit in a positive logic system, a negative voltage must be used for Vs, and the diodes reversed. The output will always be one or the other of the two input voltages. Since inverter circuits (NOT gates) cannot be made with diodes, a simple circuit using a transistor can be used.

TRIGGERED CIRCUITS

When an externally applied signal is used to bring about an immediate change in the operating state of a transistor circuit, the circuit is said to be triggered. Such circuits may be

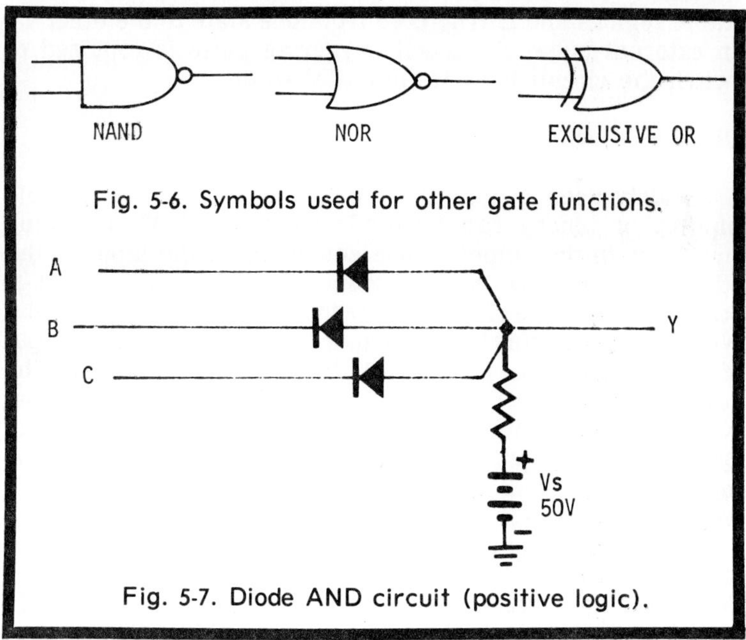

Fig. 5-6. Symbols used for other gate functions.

Fig. 5-7. Diode AND circuit (positive logic).

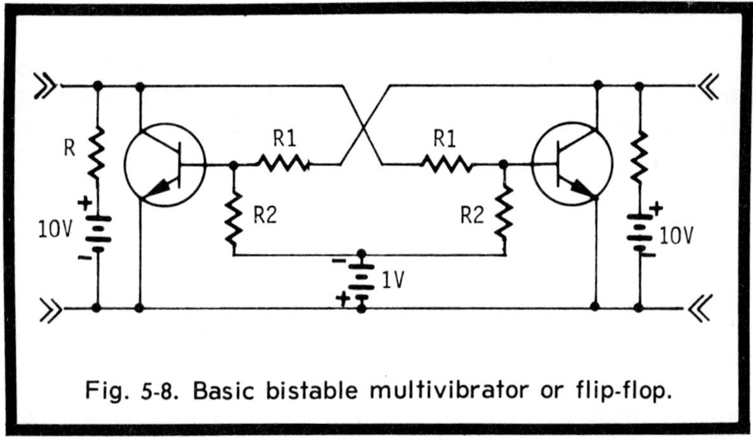

Fig. 5-8. Basic bistable multivibrator or flip-flop.

monostable, bistable, or astable. Astable triggered circuits have no stable state; they operate in the linear active region of the transistor and produce relaxation-type oscillations. A monostable circuit has one stable state in either of the stable regions (saturation or cut off); an external pulse triggers the transistor to the other stable region, but the circuit then switches back to its original state unassisted, delayed by the time constant of the circuit components. A bistable circuit (regularly called a flip-flop) has a stable state in each of two stable regions and is triggered from one state to the other by an external pulse. A second triggering pulse is required to switch the circuit back to its initial state.

Bistable Multivibrator

Widely used in digital computer circuitry, the bistable flip-flop or binary can be made from two NOT switching circuits with the output of one connected to the input of the other and vice versa. Such a combination is shown in Fig. 5-8. Starting at the base of Q1, assume that the input voltage is 10v. With the input at this level, the output is zero. Since the output of Q1 connects to the input of Q2, the base of Q2 is at zero. The voltage at the output of Q2 is at 10v, and supplies the original 10v input to Q1 which was assumed. Therefore, the output of Q2 supplies the input to Q1, and the output of Q1 provides the input to Q2. Q1 is on with zero voltage output, while Q2 is off with 10 volts output.

States other than those shown in Fig. 5-8 are possible. For example, shorting the base (input) to ground for an instant drops voltage to zero and Q1 switches off. Now we have a Q1 output of 10v; therefore, with an input of 10v Q2 turns on, providing an output of zero. So Q2 supplies the 0v input for transistor Q1, and the transistors remain in this state until the circuit is triggered again.

By causing the off transistor to turn on, or the on transistor to turn off, the circuit is forced to switch again. Once the flip-flop has switched, it remains so until either is triggered in some way. The bistable multi has only two possible states, and when one transistor flips, the other flops, which is how the circuit received its popular name. One must be on while the other is off, and by causing either to change states, the other changes of its own accord to the opposite or other state. When a transistor switch is on, the output is 0 and binary 0 has been stored; switching the transistor to off, the output is 10v and binary 1 is stored.

Triggering the Binary

In order for the flip-flop to be useful in computer and switching circuit applications, it is always necessary for it to be flipped from one state to another by means of voltage pulses. Fig. 5-9 illustrates how a binary can be triggered with a pulse applied to either collector. In applying the trigger, it is usually more desirable to turn the transistor off rather than on. The off transistor typically has a reverse-biased emitter, and this voltage must be overcome by the trigger before switching begins. The triggering pulse at -5v is applied to the Q2 collector through a capacitor. Since Q2 is off, there is a potential of +10v at the collector, which is applied to the base of Q1, holding that transistor on. When the -5v pulse is applied, the trigger input terminal voltage drops from 0 to -5v, which is passed by capacitor C, with a steep-fronted pulse, dropping the Q2 collector to +5v. Naturally, the base voltage of Q1 is reduced, and Q1 is no longer fully on. As a result, the Q1 collector voltage rises to about +6v from its former 0v; therefore, +6v is applied to the base of Q2, turning that transistor completely on. With Q2 on, its collector drops to 0v and Q1 is fully off. The circuit remains in this state until triggered again.

Triggering pulses are normally coupled to the flip-flop through a capacitor. High frequency triggering necessitates low capacitor values, with lower limits restricted by base charge considerations. As the trigger voltage reaches the coupling capacitor, the resulting trigger current forces the charge on that capacitor to change, and when the capacitor change is equivalent to the base charge, it turns the transistor off. Only partial turn off may result if either capacitor or trigger voltage is too small.

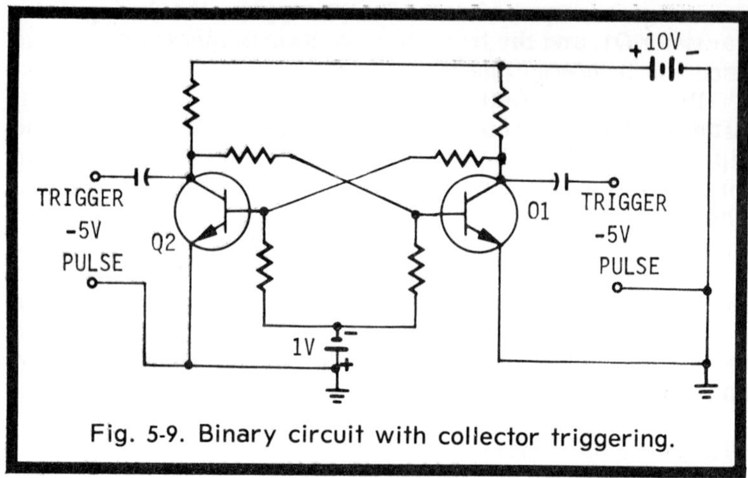

Fig. 5-9. Binary circuit with collector triggering.

Lower limits of trigger power needs can be ascertained by due consideration of the base charge requirements for maintaining the collector current in the on transistor. Circuits using high-speed switching transistors at reasonably low collector currents have minimum triggering power demands. In other words, the use of high-speed transistors in slower circuitry would be advantageous where trigger power is a consideration.

Regeneration

The introduction of the trigger pulse is all that is required to cause the on transistor to turn off and the off unit on. A pulse that causes the output voltage of the off transistor to drop below a certain point starts the action known as regeneration, with the circuit completing the switching unassisted. In Fig. 5-9, notice that the 5-volt drop at the collector of Q2 was applied to the base of Q1 and amplified by Q1, producing a rise in its collector voltage. This rise, in turn, when applied to the base of Q2, completes the switching by turning that transistor on completely.

Switching action is considerably improved with the addition of small capacitors shunted across the coupling resistors as shown in Fig. 5-10. These capacitors speed up the switching action by permitting the collector voltage change to be transferred to the opposite transistor base without loss. The coupling resistors attenuate the voltage jump, but the capacitors speed up switching time by permitting a stronger pulse to reach the base quickly. When the switching is completed the capacitors have no effect on the circuit whatever, since only DC is involved.

120

Base triggering is shown in Fig. 5-10. The pulse is capacitively coupled to either transistor in this version of the bistable multi. Applying a negative trigger to the base of Q1 turns it off, and the resulting rise in collector voltage is coupled to the base of Q2, turning Q2 on. Thus regeneration has started and the action is complete when Q1 is completely off and Q2 is fully on.

Triggering Methods

There are only two ways to produce switching action in the flip-flop. We must either trigger the on transistor off and turn the other on, or turn the off transistor on and the opposite unit off. A negative triggering pulse, when applied to the base of Q1 in Fig. 5-10, reduces the base-emitter bias, causing the transistor to begin turn-off. However, if Q1 was already off, the negative trigger would have no effect because the reverse-bias condition would exist and adding to it would only increase the reverse-bias. Although the negative pulse will not turn the off transistor on, it will turn the on semiconductor off when applied to the base. If the base of the off transistor is triggered with a positive pulse, it will begin to turn on, provided that the amplitude is greater than the reverse-bias, but it would have no effect if the transistor already was in the on state. Even though the positive pulse can flip the binary by turning the off transistor on, such action lacks the sensitivity, speed, and reliability so essential in computer switching circuits. So, for the sake of emphasis, practically all circuits are designed to

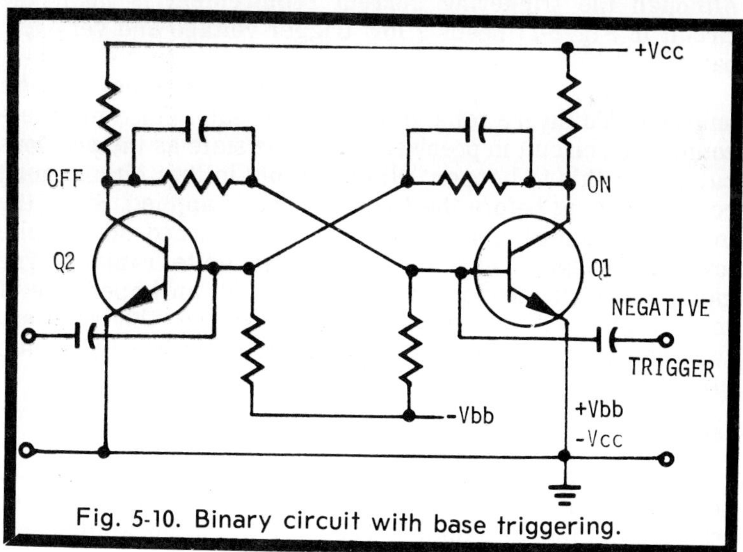

Fig. 5-10. Binary circuit with base triggering.

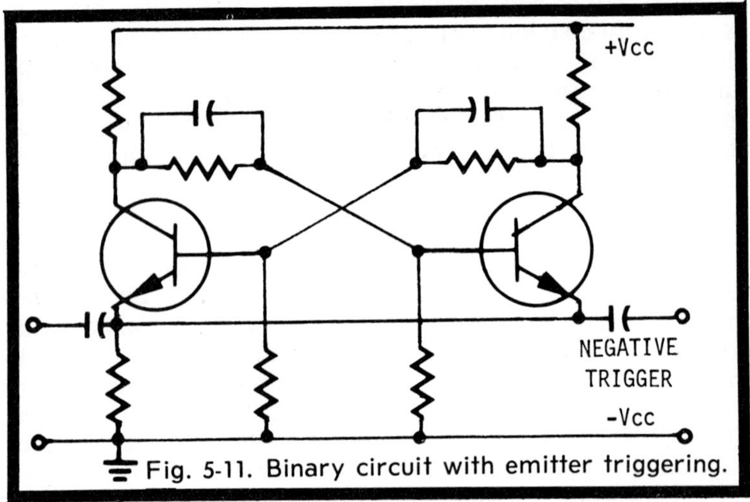

Fig. 5-11. Binary circuit with emitter triggering.

apply the triggering pulse to the **on** transistor, thus turning it off.

It is often necessary to make opposite sides of the flip-flop conduct on alternate triggering pulses, in shift registers, ring counters, etc., for example. The use of steering circuits permits this type of operation. At low-speeds, triggering at the emitters with at least a 4v pulse permits a maximum rate in excess of 2 MHz. The cross coupling network's time constant requires the use of short pulses for reliable performance. Although the triggering current requirement is high, the circuit in Fig. 5-11 needs a low trigger voltage and very few parts.

The effect of trigger pulse repetition frequency can be analyzed. To insure reliable triggering, each pulse must encounter the circuit in precisely the same state as the previous pulse. For this to happen, all capacitors in that circuit must cease charging before the trigger pulse is applied. Since the initial impedance of the transistor being turned off is quite low, the trigger capacitor discharges quite rapidly. The capacitor must recover its charge through an impedance of much higher value, and as may be assumed, this recovery time constant may limit the maximum pulse repetition rate attainable.

Isolating the Trigger

Since it may be necessary to apply trigger pulses to several circuits at the same time, collector triggering of **on** transistors must be avoided. This type of circuit would short the pulse to ground, and the pulse would not be of sufficient

122

amplitude to trigger other circuits. So called diode steering circuits permit the collector circuit of the **on** transistor to be isolated during that period.

In Fig. 5-12 the collectors are triggered with a negative pulse. Since the diode conducts during triggering, the trigger pulse is loaded by the collector load resistor. When triggering is completed, capacitor Ct recovers through biasing resistor Rt. In order to minimize trigger loading, Rt must be large; to aid recovery, it should be small. The problem may be resolved by replacing Rt with a diode, connecting the cathode to the +6v bus (Vcc). The diode's low forward resistance insures fast recovery, and its high reverse impedance prevents shunting the trigger pulse during that period. Even though large amplitude pulses are required for collector triggering, permissible variations in pulse amplitude are quite wide in most instances.

Base triggering produces steering in a way similar to collector, the basic difference being quantitative, with less energy required and more accurate control of amplitude.

Unsymmetrical Triggering Problems

Trigger pulses must be tailored to the circuit design. Pulses too short in duration or insufficient in amplitude could

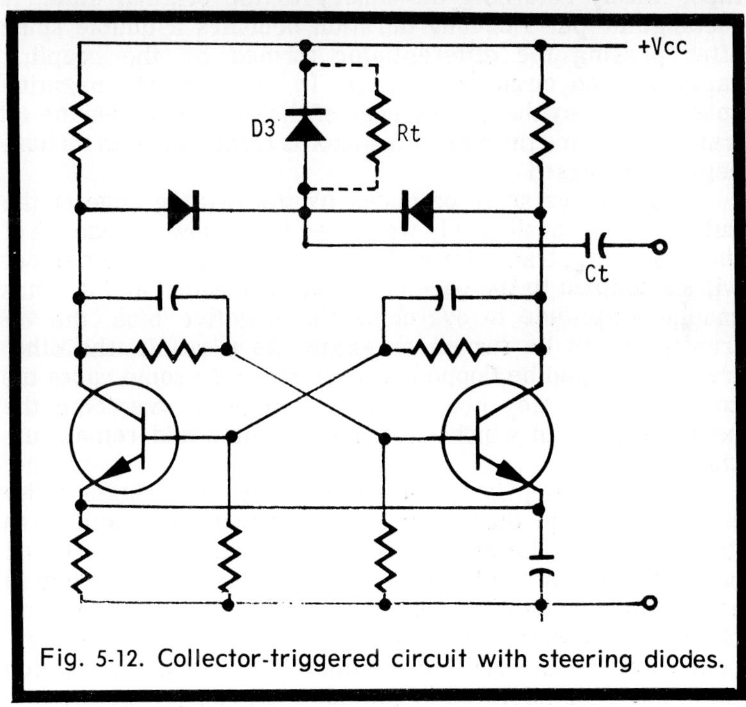

Fig. 5-12. Collector-triggered circuit with steering diodes.

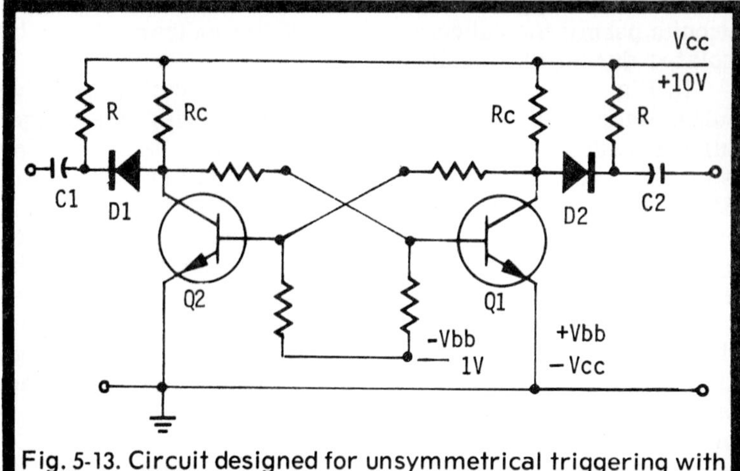

Fig. 5-13. Circuit designed for unsymmetrical triggering with isolation diodes.

result in failure to introduce enough energy into the binary to start the flipping action. On the other hand, if the pulse is too high in amplitude or too long in duration, we may acquire two flips, falsely returning the binary to the original state. A rectangular pulse of long duration becomes a double spike after passing the differentiator formed by the coupling capacitor and circuit resistance. In this case the negative spike formed by the leading edge of the pulse switches the **off** transistor **on** and the other transistor is turned **off** with its base negatively biased.

The positive spike produced by the trailing edge of the pulse may be considerably weakened by the low impedance of the conducting transistor collector circuit, but the remainder will be coupled to the base of the **off** transistor, and if it has enough amplitude to overcome the negative bias, the **off** transistor will be turned **on** again. As a result, the other transistor would be flopped **off** once more. In some cases the positive spike may not be large enough to overcome the negative bias; in which case, the circuit would remain unchanged.

If the flip-flop is operating properly, the trigger pulse should not change the circuit when the pulse is first applied to the collector of the **on** transistor. The negative spike would have no effect on the base of the **off** transistor, other than to add to the reverse-bias, but if the pulse is sufficient in amplitude, the positive spike could trigger the circuit as mentioned before. In order to eliminate this possibility, which could never be tolerated in computer circuitry regardless of

124

how small the chances are the trigger pulse amplitude can be
adjusted until satisfactory operation is assured. By connecting
large resistors in series with the trigger inputs, the possibility
of triggering on the trailing edge is greatly reduced.

Isolation diodes may also be used in the input circuit to
completely eliminate the chance of triggering errors, such as
double triggering or triggering accidentally. Fig. 5-13 shows
the application of isolation diodes in the trigger input circuit.
With Q2 on, D1 is reverse-biased by 10v. Now, as long as
neither spike (negative or positive) of the trigger coming in at
C1 is greater than 10v, the pulse is blocked by the reverse-
biased diode and cannot trigger the circuit. But if Q2 is in the
off state when the pulse arrives at C1, the negative spike
(leading edge) will trigger the circuit. So Q2 is now in the on
state and when the positive spike (trailing edge) arrives, it is
blocked.

Symmetrical Triggering

Many of the problems mentioned under unsymmetrical
triggering are applicable to symmetrical triggering, in addition to some new ones. In basic symmetrical triggering, the
trigger pulse is applied to both transistors at the same time.
With a negative pulse applied, the base current to the on
transistor is reduced, and switching action tries to begin. This
same negative pulse tries to drive the base of the off transistor
further negative, making it even more difficult for it to switch
to the conducting state. From this problem, we may assume
that successful switching without diodes requires a triggering

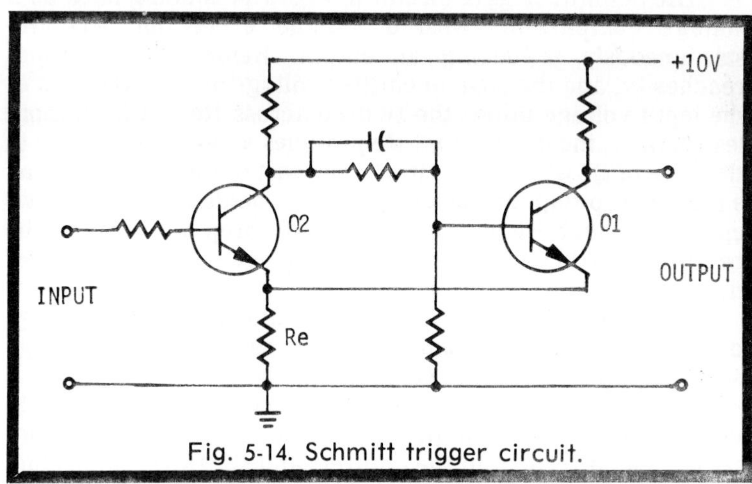

Fig. 5-14. Schmitt trigger circuit.

pulse of exactly the right amplitude and duration, as well as large speed-up capacitors.

As most of the problems normally encountered may be eliminated, or at least reduced to a tolerable level by the use of simple diodes, the circuit in Fig. 5-12 seems to provide the necessary answers. Fast operation is assured, as the low resistance of diode D3 quickly discharges the positive charge on Ct. Without the use of diodes, the major portion of the triggering pulse may be lost through the low impedance of the conducting transistor, and if the source of the trigger cannot take care of this energy loss, switching will not occur at all.

Schmitt Trigger Circuit

The Schmitt trigger is simply a regenerative bistable circuit whose state is dependent on the amplitude of the input voltage. The regenerative triggering action that takes place in a binary is put to good use in the Schmitt circuit. The circuit makes pulses with very fast rise times out of slow rise-time signals, and it is frequently used for converting sine waves to square waves. In addition, its usefulness may include signal level shifting, squaring sinusoidal or non-rectangular inputs, waveform restoration, and DC level detection.

The Schmitt trigger circuit is shown in Fig. 5-14 and bears some basic similarity to the self-biased flip-flop. The resistors supplying input to the base of Q2 were removed and the base of Q2 has a separate input terminal in the Schmitt circuit. Speed-up capacitor C1 has been added. The circuit, although still actually a flip-flop, now obtains its Q2 input externally instead of from Q1.

The Schmitt trigger circuit in Fig. 5-14 is being used as a voltage comparator, with a voltage exceeding +1v, instantaneously producing an output. Before T1, the input reaches 1v, and the base-to-emitter voltage of Q2 is the sum of the input voltage minus the 1v drop across Re. With an input less than 1v, the drop across Re provides a negative voltage to the base of Q2 which holds it off. The rest of the circuit acts as a regular flip-flop, and since Q2 is off, it holds Q1 on. Q1 draws current through Re and provides the 1v drop available in the beginning. With Q1 on, Re drops its voltage across the output, making it +1v.

The input exceeds 1v at t equals 1 and overcomes voltage drop Re. Q2 begins to conduct slightly as its collector voltage starts dropping from 10v. Q1 is then no longer held completely on, and the Q1 current through Re decreases. As the voltage drop across Re gets smaller, the base of Q2 becomes more positive with respect to the emitter, and Q2 starts to conduct

126

more heavily. Regeneration is now beginning. Very quickly Q2 is fully on and Q1 completely off, since its output voltage rises to 10v. It may be noted that the decrease in current through Re, as a result of the decrease in current through Q1, is partly remunerated by the current which flows through Q2; in other words, the current flow through R2 adds to that through Re. However, since the Q2 current is amplified by Q1 (Q1 has a gain of 50), any increase in current through Q2 results in a proportional decrease in current through Q1; that is, the current decrease through Q1 is 50 times as great as the current increase through Q2. This explains why the net current through Re is only slightly affected by the Q2 current.

As the input voltage overcomes the 1v drop through Re, the output voltage rises from 1v to 10v. Regenerative triggering is responsible for the rapid jump in output voltage level. As Q1 was turned off by Q2, it assisted in turning Q2 on by reducing the voltage drop across Re.

As the input voltage drops back to +1v, the Q2 collector current begins to decrease. As a result, the rise in collector voltage is transferred through C1 to the base of Q1, so now Q1 starts to conduct. This increases the Re voltage drop, further lowering base-emitter bias on Q2. This regenerative action rapidly flips the circuit and Q1 is on once more and Q2 off again.

To summarize, the Schmitt trigger switches Q2 on and Q1 off when the increasing input voltage equals the voltage across Re; it remains in this state as long as the input voltage exceeds 1v. As it drops back below 1v, the binary is triggered again, so Q1 is on and Q2 off. The Schmitt is frequently used as a squaring circuit.

PULSE CIRCUITS AT WORK

The basic idea in any switching circuit is a distinct change of state, such as a voltage change or a current change, or even both. The change may be utilized in the performance of logical operations, as in the digital computer, or simply to transfer energy as is the case in switching regulators. Any switch offers an extremely low resistance when closed and a very high resistance, approaching infinity, when open.

Transistor switches present the advantage of having no moving parts subject to wear or malfunction, and ease of control by various electrical inputs. Numerous functions may be performed by these semiconductor switches in varied circuitry, such as amplifiers, frequency dividers, pulse generators, and inverters, to mention a few. Applications usually characteristic of nonlinear transistor operation provide triggering, gating, and signal routing functions.

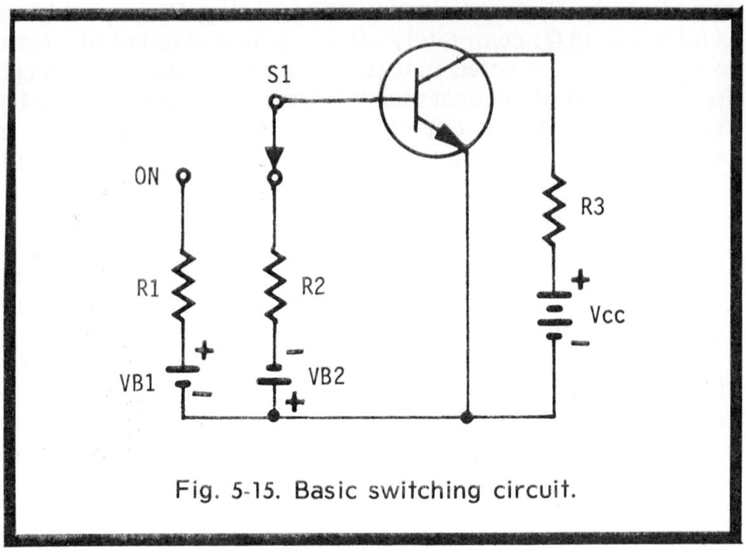

Fig. 5-15. Basic switching circuit.

With the transistor switching circuit in the **on** state, the resistance across the transistor must be very low in order to prevent a loss of power through the switch. This low resistance requirement can be met by operating the transistor in the saturation region. Ample base current to assure saturation under worst-case operating conditions must be supplied. (Worst-case design guarantees reliable operation of a circuit, even under the most unfavorable conditions.) Worst-case design considerations include capacitor, resistor, and voltage tolerances, variations in transistor parameters, and temperature effects. Just as transistor switching demands low resistance when **on**, it must be as high as possible when **off**.

Large signal operation causes a transistor to perform as an overdriven amplifier, driven from the cut-off region to the saturation region. An extremely simple switching circuit is shown in Fig. 5-15. The collector-base junction is reverse-biased through resistor R3 by Vcc. Polarity and the amount of base current are controlled by switch S1. Reverse-biasing of the emitter-base junction is provided through limiting resistor R2 when S1 is in the off position and the transistor is in the **off** or non-conducting state. The quiescent operating point refers to the value of voltage or current characteristic of the transistor circuit with no pulse in or out. Forward-bias is applied to the base-emitter junction through current-limiting resistor R1 when switch S1 is in the on position. Base and collector currents increase quickly until transistor saturation is reached. At this point Ic equals Vcc/R3.

Switching

As further increases in base drive provide no Ic increase, saturation has been reached. The active linear region is referred to as the transition region in switching operations, since the signal passes through it quite rapidly. When the collector current is maximum and the collector voltage minimum, the saturation region has been attained. This collector voltage level is known as the saturation voltage and is an important characteristic of the semiconductor. A transistor is conducting during its operation in the saturation region, and is in the **on** state, with both junctions forward-biased.

All transistor switch configurations display similar regions of operation. By reverse-biasing both junctions of the transistor, the output current is extremely small and output voltage high (cut-off). But forward-bias results in high output current and low output voltage (saturation). Now, for all practical purposes, we may completely ignore the small output current in the cut-off condition, and the small output voltage in the saturated condition.

Switching Time

Operating the switch in Fig. 5-15 from **off** to **on** and then **off**, results in the current pulses illustrated in Fig. 5-16, with a rectangular input pulse (Ib) driving the transistor from cut-off to saturation and back to cut-off. Pulse Ic representing the output current, shows distortion because of the inability of the transistor to respond instantaneously to the signal level change. The transistor response during the rise time (Tr) and fall time (Tf) is known as transient response. The length of time the transistor remains cut off after the input pulse is applied is Td or delay time. This finite time is necessary

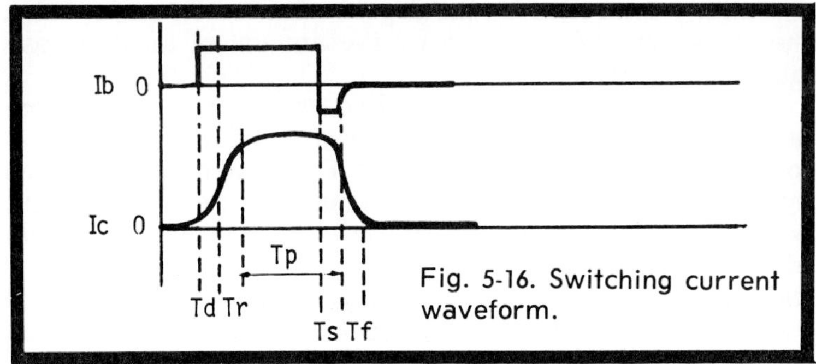

Fig. 5-16. Switching current waveform.

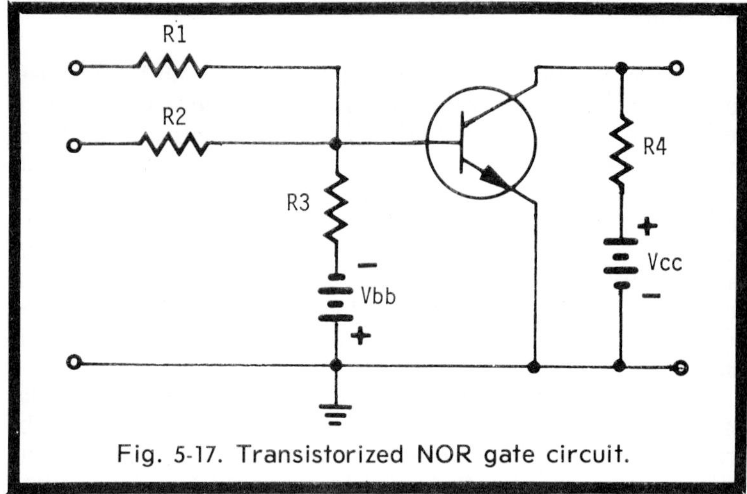

Fig. 5-17. Transistorized NOR gate circuit.

before the applied forward-bias overcomes the emitter capacitance of the transistor and collector current starts to flow.

The build-up time Tr (rise time) is the time required for the leading edge of the pulse to increase in amplitude from 10 to 90 percent of maximum value. By overdriving the transistor, it is possible to reduce the rise time, but only small amounts of overdrive may be used, since turn-off time is affected as a result. The storage time plus the fall time adds up to the turn-off time.

The pulse time Tp or pulse duration is the length of time the pulse remains at, or in the proximity of, its maximum value. Pulse duration is measured between the points on the leading edge and the trailing edge showing an amplitude of 90 percent of maximum value. Storage time (Ts) is the length of time that output current Ic remains at its maximum value after input current Ib is reversed. The length of storage time is essentially controlled by the degree of saturation into which the transistor is driven and by the amount of turn-off base current available.

The fall time (Tf), often called decay time, is the time needed for the trailing edge to decrease in amplitude from 90 to 10 percent of its maximum value. Fall time may be reduced by the application of a reverse current at the end of the input pulse. The total turn-on time of a transistor switch is the sum of the delay time and the rise time. The total turn-off time is the sum of the storage time and the fall time. A reduction in either fall time or storage time decreases turn-off time and increases the potential pulse repetition rate of the circuit.

130

A transistor switching circuit operating as a typical open or short circuit forms a gate. Such circuits are used quite often in computer applications to perform a number of functions; i.e., circuit triggering at predetermined intervals, and level and waveform control. Since such circuits are designed to evaluate input conditions to produce a prescribed output, their primary uses are in logic applications. Logic circuits such as AND, OR, NAND, NOR, clamping, and shunt or inhibitor circuits are used extensively in computers.

The OR gate has only one output, but must have two or more inputs. It offers a prescribed output condition when one or another input condition exists. With a properly polarized pulse applied to one or more of the inputs of an OR gate, an output pulse of the same polarity is obtained. When phase inversion of the input signal is required, the OR gate is changed to a NOR gate. We arrive at the NOR label from a contraction of NOT OR, because it produces, in effect, the opposite state of the OR gate. Fig. 5-17 shows the transistor version of the NOR gate. Bias values must be chosen so the transistor is off when all inputs are low and on with either or both of the inputs high.

The AND gate also has only one output and two or more inputs. It offers an output when all inputs are applied at the same time. Like the OR gate, by using a configuration that provides phase inversion, the NAND gate is produced. A direct-coupled transistor logic circuit used to trigger a bistable multivibrator, the overall gating function consisting of a NAND and a NOR, is shown in Fig. 5-18. Series connected

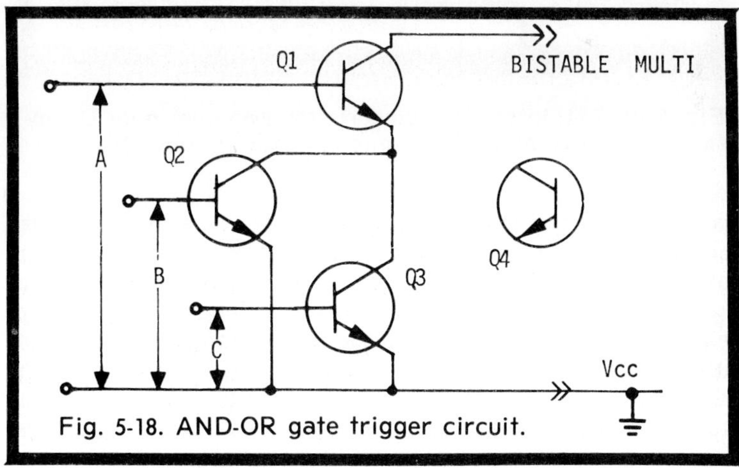

Fig. 5-18. AND-OR gate trigger circuit.

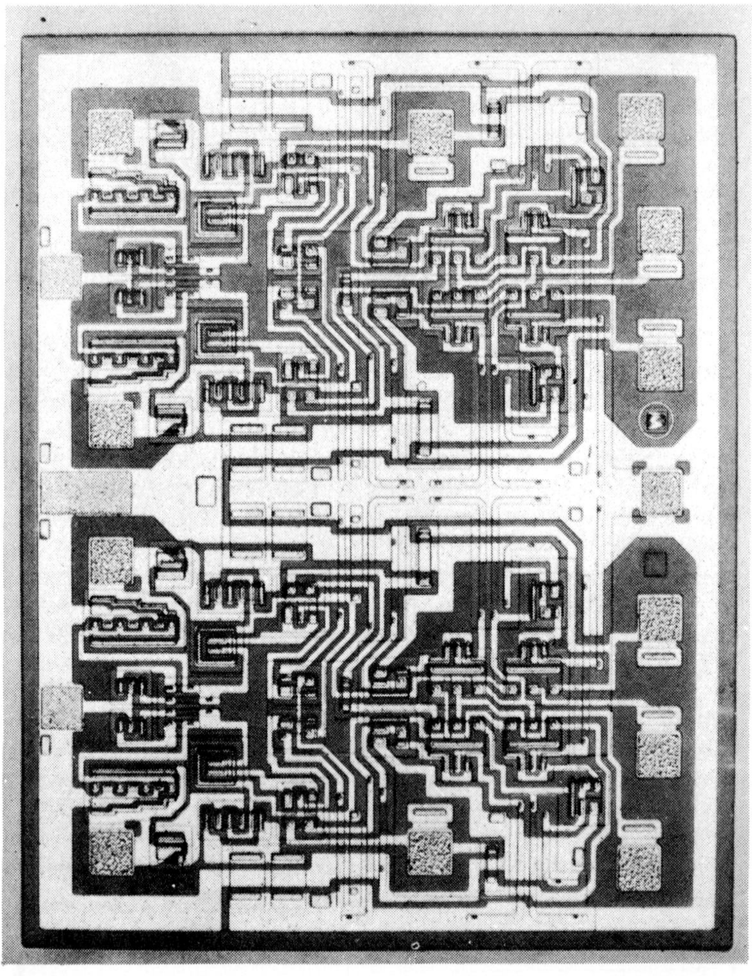

Fig. 5-19. TTL integrated circuit; MC3060 dual type D flip-flop. (Courtesy Motorola Semiconductor Products)

transistors Q1 and Q2 form a NAND gate; Q1 and Q3 are also series connected, forming a NAND gate. Q2 and Q3 are parallel connected and thus form a NOR gate. Without input signals all transistors are cut off, but a triggering pulse appears at the Q1 collector when the proper input conditions for either NAND gate are met. Q1 and Q2, or Q2 and Q3, must be triggered into conduction. Gating circuits can be used as amplitude discriminators, better known as limiters, clippers, clampers, and signal-shunting or tranmission gates.

Propagation Delay

Propagation delay is the most important consideration in ascertaining a logic system's speed capabilities, so essential to computer applications. Since delay time does limit the maximum speed with which information may be processed in a computer, typical propagation delays ranging from several microseconds to less than a few nanoseconds can be achieved, depending upon the transistor and circuit used.

RTL LOGIC

RTL (resistance-transistor logic) is the simplest of all computer building blocks, performing a NOR function if positive voltage levels are defined as binary 1 and negative voltages are designated as binary 0. The design of RTL circuits demands DC stability under worst-case conditions. If optimum switching performance is desired, the circuits should be designed to provide maximum reverse base current for a specific fan-in (number of inputs) and fan-out (number of outputs). Storage and fall times are decreased by this approach; thus, smaller propagation delays are provided per stage, but the fan-out capability of the circuit is decreased.

RTL circuit propagation delay measurements are made under worst-case conditions; that is, alternate stages are subjected in turn to maximum and then minimum drive conditions. Maximum drive provides short delay and rise times but long storage and fall times; maximum drive occurs when a given stage is driven by three unloaded stages. Minimum drive produces short storage and fall times but long delay and rise times; minimum drive occurs when a given stage is driven at only one input by a fully loaded stage.

TTL LOGIC

A TTL (transistor-transistor logic) circuit is a type of dual flip-flop that triggers on the positive edge of the clock and performs the type D flip-flop logic function. The integrated circuit illustrated in Fig. 5-19 consists of two completely independent Type D flip-flops, both having direct $\overline{\text{SET}}$ and $\overline{\text{RESET}}$ inputs for asynchronous operations such as parallel data entry in shift register applications.

Information may be applied to, or changed at, the D inputs at any time during the clock cycle except during the time interval between the setup and hold times. The inputs are inhibited when the clock pulse is high and data may be applied to the input steering section of the flip-flop when the clock pulse goes low. The information present at the inputs during

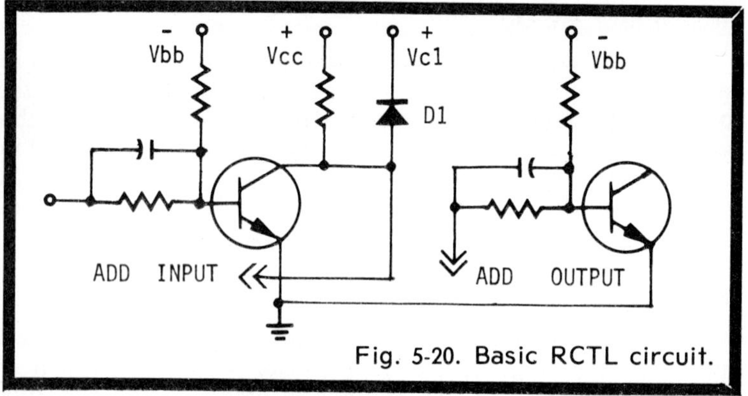

Fig. 5-20. Basic RCTL circuit.

the time interval between the setup and hold times is transferred to the bistable section on the positive edge of the clock, and the outputs Q and $\overline{Q}$ respond accordingly. The MC3160 flip-flop can also be set directly at any time, regardless of the state of the clock, by applying a low state to the direct $\overline{SET}$ or $\overline{RESET}$ inputs.

RCTL LOGIC

The RCTL circuit (resistance-capacitance-transistor logic) uses many transistors as a rule and offers the capability of very fast operation. The logic function provided by the RCTL configuration is quite similar to the RTL previously described. An RCTL circuit is shown in Fig. 5-20.

As result of placing speed-up capacitors across the coupling resistors, the RCTL unit displays the desirable feature of exceptionally fast operation. Aside from compensating for the stored energy in the transistor, the capacitor produces a sizable forward base current on an instantaneous surge as a result of the low-resistance path offered to the pulse energy. Since very-high-speed switching times are available, the maximum repetition rate is limited by the speed-up capacitor's value. This makes it essential to choose the proper value capacitor, just large enough to balance the stored charge of the transistor but not large enough to affect the switching rate.

DTL LOGIC

Diode-transistor logic circuits may perform a NAND or a NOR function as shown in Fig. 5-21. The definition of the voltage levels is the determining factor, and, as may be anticipated, the circuit is capable of extremely high speed. It uses an arrangement of many diodes and comparatively few

134

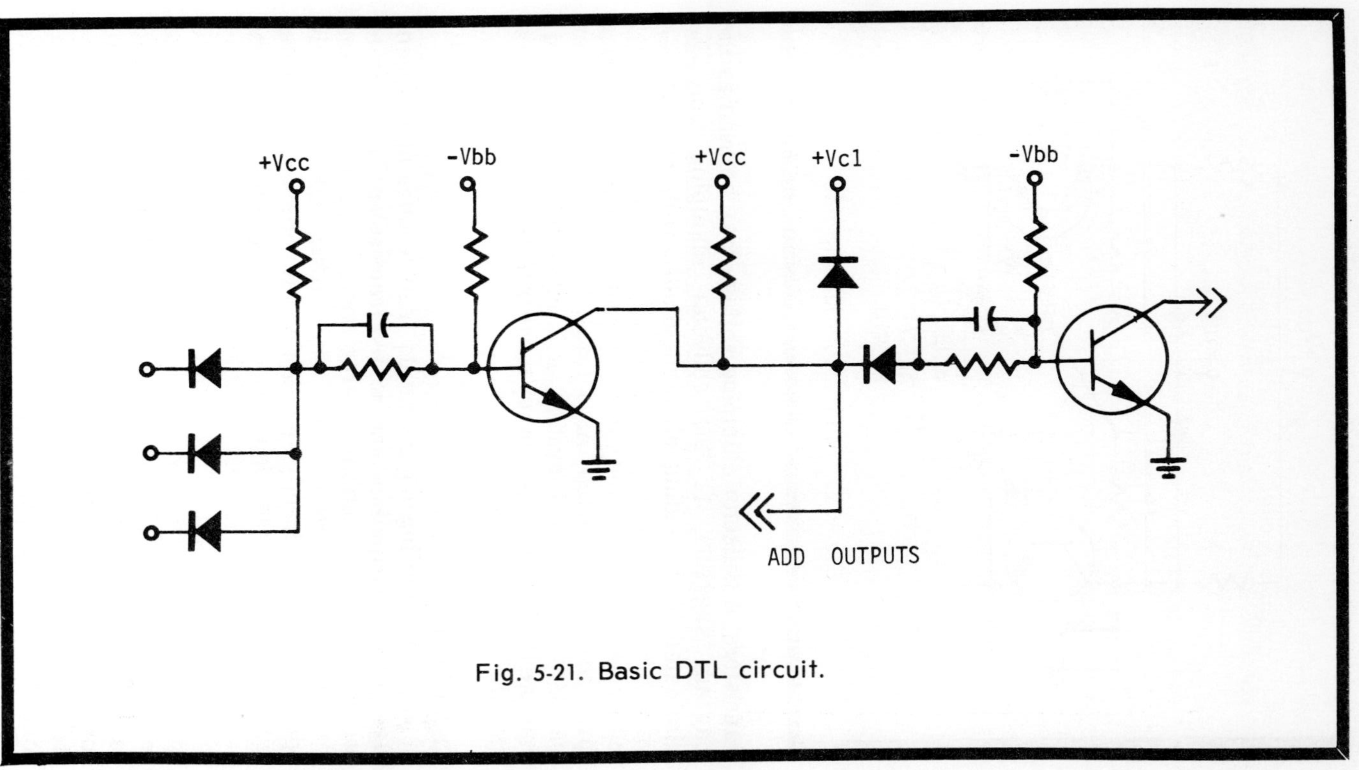

Fig. 5-21. Basic DTL circuit.

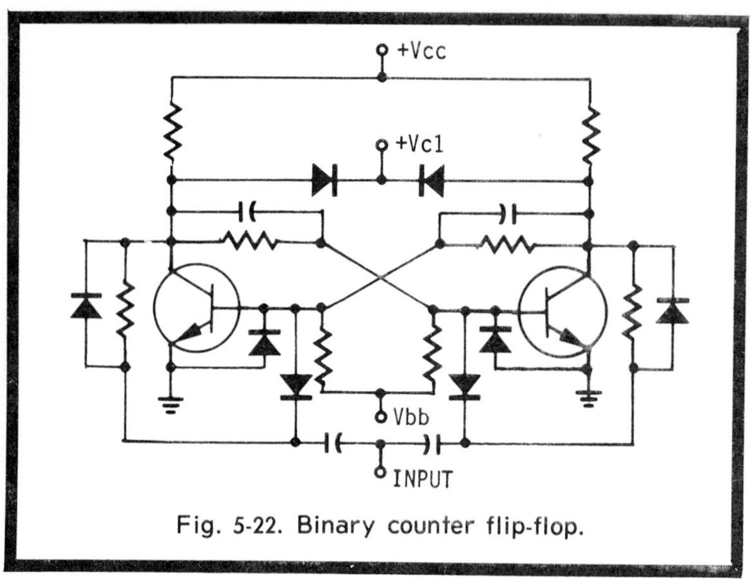

Fig. 5-22. Binary counter flip-flop.

transistors. A collector clamping voltage may be used as the diagram suggests, or the collector clamping may be eliminated by providing all input diodes with reverse-bias when the transistor is **on**. Although the latter arrangement makes larger fan-in and fan-out possible, somewhat slower speed may be expected. As a result of the need for fewer transistors in the DTL system, it is much more economical than the RCTL which uses many more transistors to perform a specific logic function.

CSL LOGIC

Current-steering logic circuits feature ultra-high speed, high power dissipation, and many transistors. In order to provide multiple functions, a large number of transistors are involved. As a result, AND and NAND, OR and NOR operations are directly available with the combination (function and inverse) dependent on voltage level definition.

The use of the smallest possible load resistor is a necessity in the proper design of current-steering circuits, since the speed is not only related to, but also limited by, the time constant of the combination of the load resistance and the load capacitance. Some degree of superiority is provided by the complementary symmetry approach over the diode current steering concept since it may be designed with less critical component tolerances. Equal speed and reliability are provided by the two approaches to the problem.

136

BINARY COUNTER

Many flip-flops are needed for the temporary storage of information in computer operation. Any of the basic logic circuits may be used to form the set-reset flip-flop. The binary counter type of flip-flop, similar in design to the RCTL system, is shown in Fig. 5-22.

The trigger gating circuit must be designed so that a negative-going pulse will turn the on transistor off when applied to the input. This means that the size of the capacitors at the input are governed by the value of the maximum stored charge of the transistor and the amplitude of the input voltage swing. As time constant problems can be a disturbing factor at high frequencies, additional diodes are connected from base to emitter of each transistor. These and the two diodes in parallel with the gating resistors will eliminate the high-frequency troubles, if encountered. However, if high-frequency operation is not proposed, the diodes are not necessary.

Noise Control

As ultra-high-speed computer system operations are contemplated, the matter of noise interference increases in significance. Noise immunity relates to the ability of a given circuit to ignore the amplitude and duration of specific noise voltages, within reasonable limits. Computer circuits, as a general rule, are subject to three primary sources of noise:

> Capacitive cross-coupling
> Inductive cross-coupling
> Common impedance coupling

The inductive noise component is usually the most important due to the low voltages and high currents inherent in the circuitry.

When considering a switching design for optimum noise immunity, it is important to determine the approximate level of the noise-voltage amplitude that would be required at the input to produce a change at the output. As this amplitude is a direct function of the transient response of the switching circuit, the pulse width or duration of the noise voltage must also be evaluated. As further consideration is given to noise effects, it is assumed that the noise voltage is of sufficient duration that effects on the circuit's transient response may be ignored; noise voltage may be no less than maximum turn-on or turn-off time in the switching circuit.

For the sake of simplicity, the DTL circuit in Fig. 5-21 provides an illustration of the design of a logic circuit for proper noise immunity. If all inputs are high, a negative noise pulse at any input would probably trigger the **on** transistor **off**, but a positive noise pulse would have no effect on the circuit. The level of noise (Vn) necessary to produce a change is determined by the reverse bias on the input diodes (Vr), the amount of forward bias (Vf) needed to produce appreciable conduction of an input diode, and the stored charge (Qs) of the **on** transistor. For the **on** state, therefore, the negative noise voltage amplitude or level necessary to bring about a change in the output is given by the formula:

$$-Vn = Vr - Vf + (Qs/Cf)$$

Should any one of the inputs be low while the transistor is in the **off** state, only a positive noise pulse introduced at a low input would have any affect whatsoever on the transistor output. The level of the positive noise voltage necessary to begin the turn-**on** of the transistor is ascertained by the reverse bias voltage on the base-to-emitter junction of the transistor (Vb), the forward bias (Vbe) needed across the base-to-emitter junction to produce appreciable conduction of base current, and the amount of charge necessary to charge the input capacitance (C1) at the base through the voltage Vb + Vbe. For the **off** condition, the positive noise voltage level is given by the formula:

$$Vn = (Vb + Vbe)\ (1 + \frac{C1}{Cf})$$

The noise immunity figure can be defined for a specific circuit as the ratio of the noise voltages determined by the formula to the voltage swing of a true input (the true input is approximately equivalent to the collector supply voltage). Having similar or equal noise immunity for both **on** and **off** conditions is desirable, since the percent noise immunity figure for any circuit is no better than the lower of the two values.

Due to the fact that values Vf, Vbe, Qs, Cf, and C1 are constants for a certain transistor and diode, the values Vr and Vb may be selected for a desired noise immunity for a given circuit design. It should be remembered that circuit noise immunity and fan-out capability are interdependent; if noise immunity is made too great, the fan-out capability will be reduced. As a result of this situation, a compromise between the two is a necessity.

MOSFET LOGIC

Definitely well-suited for digital-type logic circuitry, enhancement-type or field-effect transistors permit switching circuits that would be out of the question from a practical standpoint with other types of semiconductors. MOS (metal-oxide-semiconductor) field-effect transistors have a metal control gate which is separated from the semiconductor channel by an insulating oxide layer. MOSFET transistors have extremely high input impedance, which is not affected by bias polarity on the gate, and leakage currents are more or less unaffected by temperature changes.

Two types of MOSFETs, the depletion type and the enhancement type, are available at this time. Although only the latter is of interest for computer work, both will be briefly discussed before getting into the many digital computer circuits made possible by the enhancement type.

Charge carriers are present in the channel of the depletion type when no bias voltage is applied to the gate. The channel conductivity is reduced when a reverse gate voltage depletes this charge. Channel conductivity is increased by a forward gate voltage which attracts more charge carriers into the channel. The enhancement type requires a forward-biased gate to produce active carriers, and thus allows conduction through the channel. With either zero or reverse gate bias, no useful channel conductivity occurs.

Although the basic types of MOSFETs number only two, both types are available in either N-channel or P-channel. This provides four distinct types of MOS field-effect transistors,

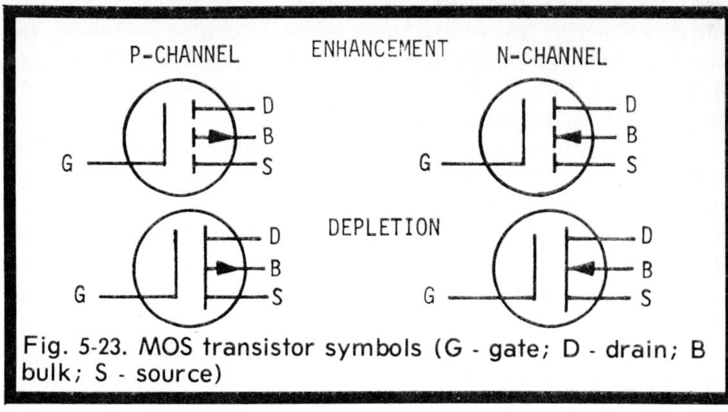

Fig. 5-23. MOS transistor symbols (G - gate; D - drain; B - bulk; S - source)

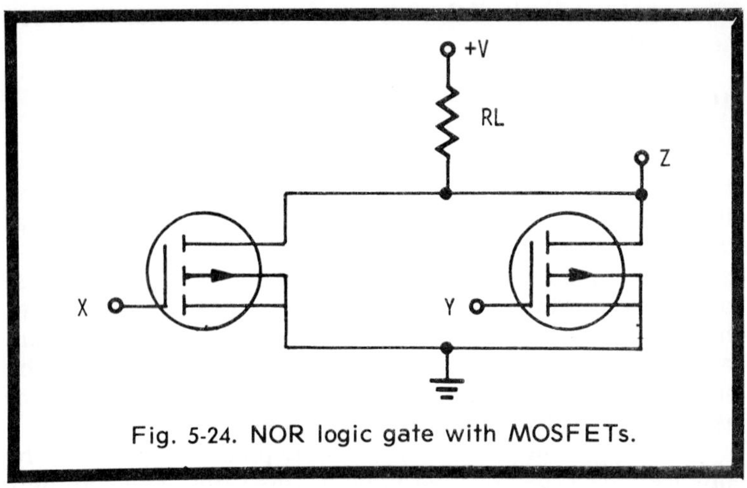

Fig. 5-24. NOR logic gate with MOSFETs.

because either electron conduction (N-channel), or hole conduction (P-channel) are utilized in depletion-type and enhancement-type units. Fig. 5-23 shows the symbols for the various MOS transistors. The direction of the arrowhead indicates the channel type; the arrow pointing out is a P-channel and pointing in is an N-channel. The solid channel line is a normally-on depletion-type transistor, while the broken line indicates a normally-off enhancement-type transistor.

When the N-channel enhancement MOS is positively biased at the gate, electrons are attracted into the region under the gate. If enough voltage is applied, the character of the region changes from P-type to N-type, producing a conduction path between the source and drain. In P-channel enhancement transistors, a negative bias voltage draws holes into the area beneath the gate, changing the channel region from N-type to P-type, and providing a conduction path from source to drain. Increased gate voltage causes a shift in the forward transfer characteristic along the gate voltage axis. This feature causes the enhancement-type transistor to be well suited for switching applications.

Direct-coupled signal inversion without shifting levels between stages is an advantage of enhancement type MOS. The relationship between the saturation voltage (Vd) and threshold voltage (Vth) is a most important consideration for MOS logic circuits. Direct coupling necessitates that Vd (sat) is smaller than Vth, and it is relatively easy to design enhancement transistors meeting this need.

A simple NOR gate with two MOS transistors and a load resistor is shown in Fig. 5-24. If the supply voltage is lower than Vth, the X and Y inputs are low; the inputs are high if the

140

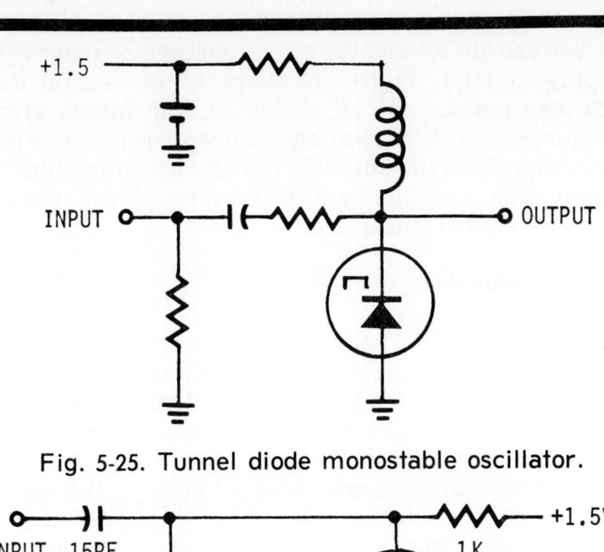

Fig. 5-25. Tunnel diode monostable oscillator.

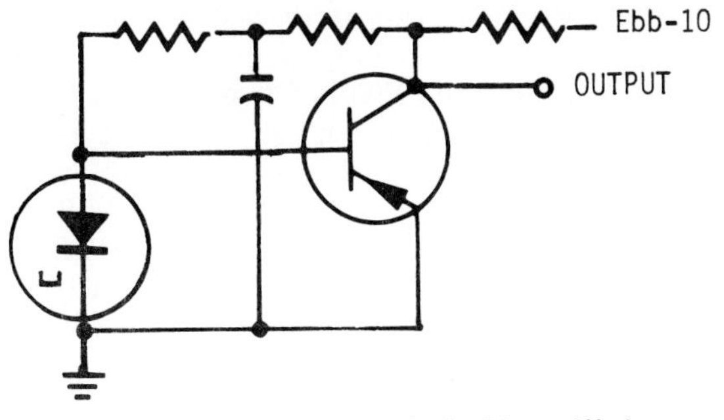

Fig. 5-26. Tunnel diode flip-flop.

Fig. 5-27. Astable hybrid oscillator.

voltage is higher than Vgo. With both inputs low, both MOS transistors are cut off and the output voltage is about equal to the supply voltage (high), as there is very little current through load resistor RL. If either or both inputs are high, current forces the output voltage to drop down to the level of Vd (sat); therefore, the output is low. By assigning binary 1 to the higher level and binary 0 to the lower level, the circuit functions as a NOR gate.

MOSFET Problems

Proper MOSFET performance requires the relative perfection of an extremely thin layer of insulating material between the gate and the channel. Should this critical layer be punctured by the accidental application of an above-normal voltage to the gate terminal, the damage cannot be corrected. However, if the damaged area is small enough, the additional leakage current introduced may be negligible for most general purposes. But extensive damage can reduce the efficiency of the unit to that of an ordinary junction gate type field-effect transistor, and leakage levels could be quite unsatisfactory. Thus the importance of appropriate precautions cannot be over-emphasized, and measures should be taken to insure compliance with manufacturer's gate-voltage ratings.

The principal danger to the delicate gate insulation is from static electricity, which strangely enough can accumulate on the gate electrode when the transistor is permitted to slip around in a plastic container, or even if the leads are rubbed against material containing silk or nylon fibers. Such hazards may be avoided by the use of metal containers, wrapping transistor leads in foil, or somehow electrically short-circuiting the leads during shipping, carrying, or storing.

Another problem arises from handling by personnel. Damage to the gate insulation may result from electrostatic charges. An average person may acquire a static charge of about 300 volts at a humidity level of 35 percent, and simply by plugging the transistor into a socket, or any type of circuit connection where the gate lead makes contact with ground before the other leads, can damage the unit. By using an electrostatic grounding strap when handling MOSFETs, this danger is eliminated. Although the grounding strap may have several megohms impedance to ground, it will permit the static charge to leak off quickly and safely.

When installed, the circuit impedance is normally low enough to protect the transistors and to prevent electrostatic accumulation. Even though risk is involved by careless handling before the MOS transistors become part of the actual circuits, thousands of hours of trouble-free operation may be anticipated thereafter under practical conditions.

TUNNEL-DIODE SWITCHING CIRCUITS

The tunnel diode is a very useful switching device, and in computer applications it offers such important advantages as small size, high speed, and low power requirements. As a result of its extremely fast switching capabilities, transition times of 27 picoseconds are possible. However, the speed of the average high-speed circuit is not limited by the tunnel diode itself, but rather by the external capacitance and inductance. Most available tunnel diodes are made from germanium or gallium arsenide, with the former featuring exceptional advantages in high-speed switching; germanium types have a rise time as low as 40 picoseconds and a very low noise figure. The gallium arsenide diode offers a much greater power-handling capability than the germanium device and outputs may normally approach four times the amplitude. This makes the gallium unit particularly attractive in the microwave oscillator field.

Two low power consumption tunnel-diode multivibrator circuits are shown in Figs. 5-25 and 5-26. The tunnel diode is also used in hybrid multivibrator circuits with considerable success, the transistor providing amplification as in Fig. 5-27. High-speed logic circuits, such as the OR gate shown in Fig. 5-28, and the AND gate in Fig. 5-29, take advantage of the extra high speed switching capability of the tunnel diode.

Switching operation is made possible by a load line that intersects the diode characteristic in three points as illustrated in Fig. 5-30. Only Points C and E are stable operating points, and if the circuit is operated at Point C with a positive current of sufficient level introduced, the operating point rapidly switches to Point E. In the same manner, a negative input pulse switches the operating point quickly to C again. One of the major advantages of the switching mode is its lack of sensitivity to the precise linearity of the negative-resistance area of the tunnel diode characteristics. Minor irregularities in the negative characteristic have an insignificant effect on the normal switching action of the tunnel diode.

The static load line for the basic monostable gate is determined by the voltage and resistance. If the Ro value is lower than the minimum dynamic negative resistance of the

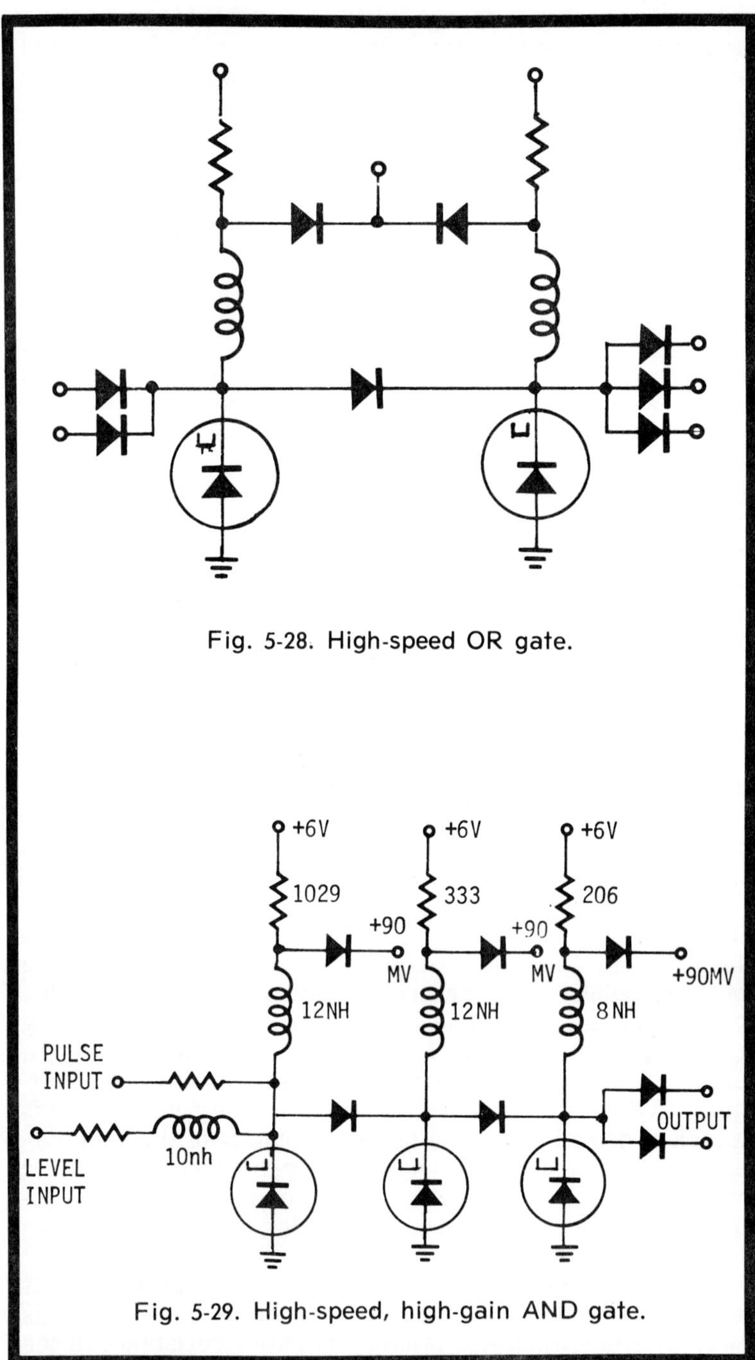

Fig. 5-28. High-speed OR gate.

Fig. 5-29. High-speed, high-gain AND gate.

tunnel diode, a single operating point is all that exists. By adjusting the voltage to bring the operating point to E, the gate is stable in its low state. The inductive time constant L/Ro determines the dynamic load line, and the current in the circuit is actually constant when the switching time is short in comparison to the inductive time constant.

By applying a small step of current across the diode, the operating point switches to F, which is the high-voltage mark along the constant-current path. Now the energy stored in inductor L must be dissipated before a return to the original operating point is possible. The energy in L decreases, and the operating point moves along the tunnel diode characteristic until minimum current is reached at Point G. Now switching again proceeds along the constant-current path to Point H, completing the cycle of operation by a recovery region in which the energy in the inductor is permitted to build up to its starting level, and operation moves during this time to its original point.

Fig. 5-31 shows a simple form of a tunnel-diode logic circuit for an AND or OR function. Adjusting the static operating bias permits the diode to be triggered by a single input, thus performing an OR function. If bias is such that all inputs are needed to trigger the tunnel diode, the AND function is obtained. Since coupling impedance is quite high compared to diode impedance, the inputs can be considered as sources of current during the triggering time. Biasing for the triple-input AND gate is shown Fig. 5-31. By increasing the operating point bias to a slightly higher level, the circuit will trigger on two of its inputs. This function would simulate that of a logic majority gate.

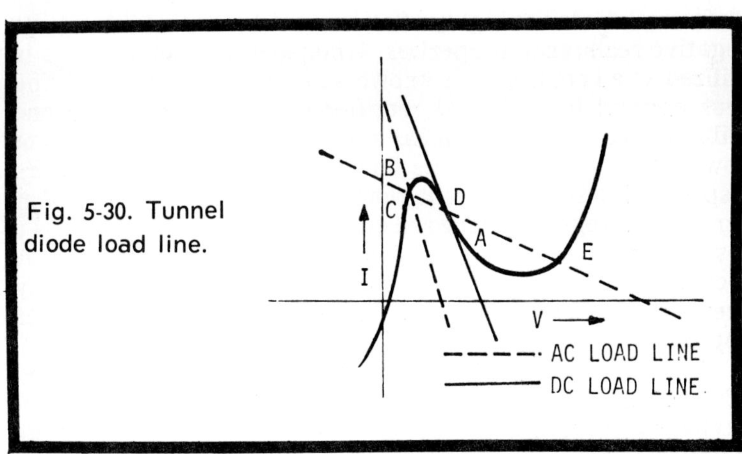

Fig. 5-30. Tunnel diode load line.

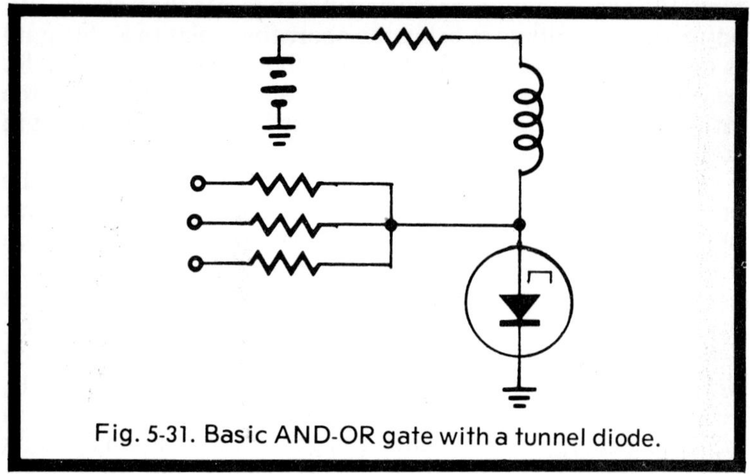

Fig. 5-31. Basic AND-OR gate with a tunnel diode.

Radiation Resistance

Although such requirements are not always prevalent, resistance to nuclear radiation can be clearly marked as a major feature of the tunnel diode. Conclusive tests have shown the tunnel group to be many times more resistant to radiation than ordinary semiconductors. As the initial resistivity of the tunnel diode is so low, it is not noticeably affected until large amounts of radiation have been applied. They are also less affected by ionizing radiation, because of their lack of sensitivity to surface changes thereby produced.

Tunnel Rectifiers

The tunnel diode is a highly efficient rectifier, in addition to its negative-resistance properties. Whenever the tunnel diode is utilized as a rectifier, it is known as a "tunnel rectifier." The peak current for a tunnel rectifier is slightly less than one milliampere. In conventional rectifiers the forward current flow is large and the reverse flow is minute. Tunnel rectifiers display substantial current flow in the reverse direction at very low voltages and forward current flow is quite small. As a result, tunnel rectifiers allow smaller signal voltages to be rectified than conventional types, although their polarity needs are reversed and the "back diode" name becomes appropriate.

Its high-speed switching capablity and exceptional rectification characteristics make the tunnel rectifier an ideal device for coupling in one direction and nearly complete isolation in the other. The use of tunnel rectifiers to provide directional coupling in a tunnel diode logic circuit of simple proportions is shown in Fig. 5-32.

Use of the tunnel diode as an oscillator, switch, amplifier, or rectifier depends upon circuit values and applied voltages. Parameters determining the mode of operation are shown in Fig. 5-33. They include series resistance, load resistance, junction capacitance, series inductance, and the negative resistance of the tunnel diode. For simplification, these may be combined into two parameters:

$$Alpha = (Rs+RL)/R$$
$$Beta = (Rs+RL)RCj/Ls$$

The value of alpha indicates whether the diode will operate as a switch. With alpha exceeding 1, the diode performs as a bistable switch since the load line intersects the I-V characteristic at two stable points (Fig. 5-33B). When the circuit is biased to operate at Point A and a positive pulse is applied, the operating point moves to Point B. When a negative pulse is applied, the circuit switches back to its original point at A. Thus the diode properly functions as a switch, and displays a capabliity needed for computer logic circuitry.

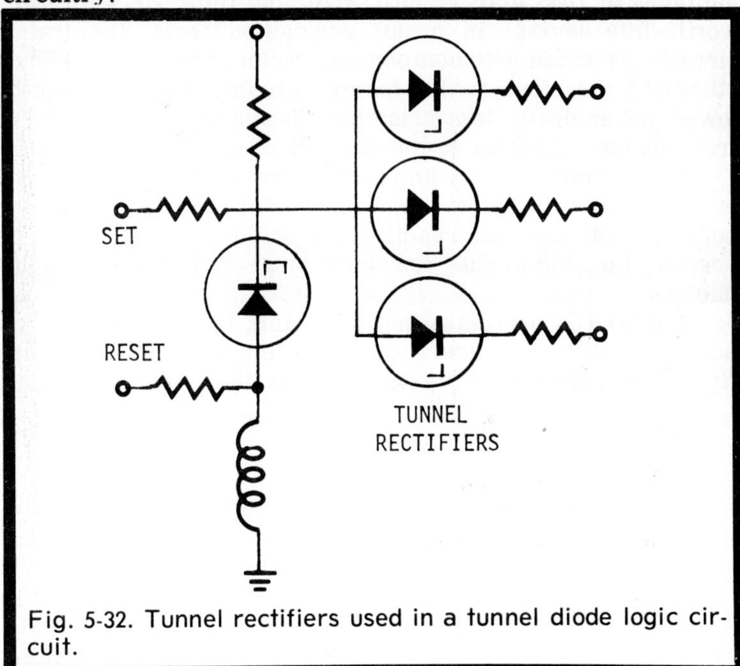

Fig. 5-32. Tunnel rectifiers used in a tunnel diode logic circuit.

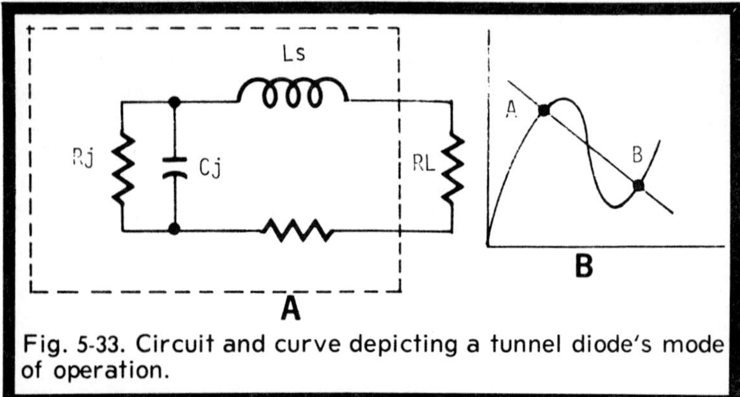

Fig. 5-33. Circuit and curve depicting a tunnel diode's mode of operation.

DIGITAL ICs

As we advance in space technology, the development of ICs has forced us to weigh their advantages against the disadvantages, and now strangely enough, we must find them worthy of favorable consideration.

There are many types of ICs available, some containing the equivalent of as many as 21 transistors and 27 resistors in a single TO-5 package. We can readily recognize three advantages at first glance: cost, size, and time. All add up to worthwhile savings in major production costs. Integrated circuits are made with components in close proximity to each other and, therefore, realize better matching, finer tolerances, lower power drain, less noise pickup and generation, and a tremendous reduction in the number of solder connections.

Replacement is easy and schematics are not complicated, even for the uninitiated. Not available at this time are components that can be readily connected externally where needed; i.e., high-value resistors, large capacitors, and inductors.

Digital ICs are classified according to purpose such as adders, drivers, expanders, gates, buffers, inverters, and flip-flops, utilizing some of the designations discussed previously:

RTL - Resistor-transistor logic

mW RTL - Low-power resistor-transistor logic

DTL - Diode-transistor logic

VTL - Variable-threshold logic

HTL - High-threshold logic

148

ECL - Emitter-coupled logic (current-mode logic)

T2L - Transistor-transistor logic

Mounting and Packaging

ICs are mounted on perforated or printed-circuit boards and connected by soldering or plug-in sockets, with the latter preferred, to eliminate strain and possible heat damage to the unit. Integrated circuits vary widely in packaging by different manufacturers, but about 90 percent use one of four types, Case 93, TO-91, TO-99 (8-pin), or TO-100 (10-pin). Low-wattage soldering irons must be used when working with ICs; 25 watts is ample and safe. Leads must be short and power supplies bypassed to ground near the IC input. Parasitic oscillations may result if lead lengths are not extremely short.

Integrated Logic

Integrated circuit design is more suitable for logic types where only two discrete signal conditions exist in digital systems, such as **on** and **off**. In the linear type, or analog circuit, the signal may have any value between full **on** and full **off**. Operating levels are fixed in logic circuitry (**on** and or **off**), and the sensitivity at these levels must be much lower than would be possible in the linear operation region. Reliable performance is assured as such circuits, being repetitions of an accepted pattern, are easier and more readily fabricated economically. As an example, by using a simple AND gate, we

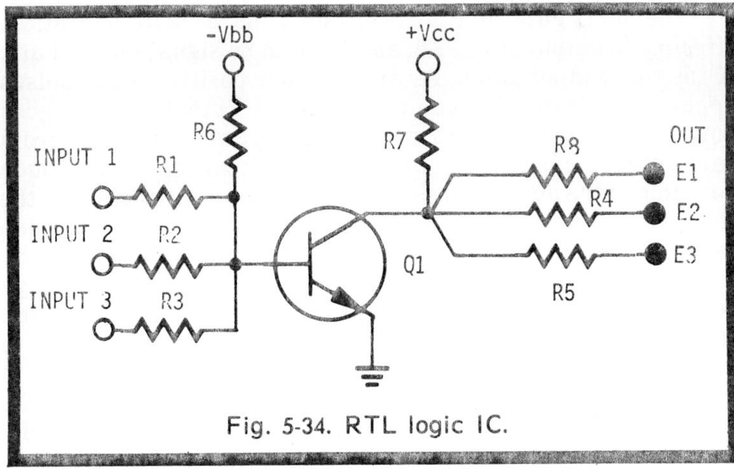

Fig. 5-34. RTL logic IC.

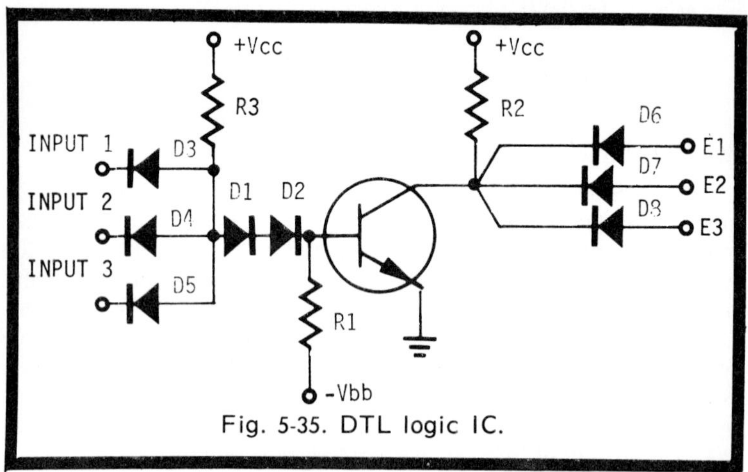

Fig. 5-35. DTL logic IC.

can perform any kind of logic function. Circuit design, regardless of complications, is based upon a single gate.

Semiconductor gates are affected less by temperature variations, a characteristic related to sensitivity considerations. When a transistor is in the **off** state, the leakage current in a planar diffused structure is less than in the linear mode of operation, and variations in leakage current will not be detected. If the transistor is **on**, the impedance level is very low and variations will be in the millivolt area. The many types of logic blocks have their particular area of superiority, and the many variables must be carefully weighed by the designer.

Typical IC Logic

The RTL gate in Fig. 5-34 performs a NOR operation, offering an output at E1, E2, and E3 with no signal input at any of the three input gates. As one or more positive input pulses appear, the transistor conducts and the output voltage goes down. Since the resistors do not have perfect isolation from each other, conduction through any resistor will induce currents in the adjacent resistors. It must be apparent that the feedback between elements, in addition to the value of each resistor, is critical in ascertaining the **on** condition of the transistor. Since resistors effect high power losses and large resistors introduce appreciable time delays in logic circuits, time delay quickly becomes a limiting factor in the speed of operation of the system–an important consideration in computer circuits. Increasing the speed of logic gates by the use of speed-up capacitors shunting the resistors is advisable.

The basic DTL integrated circuit is shown in Fig. 5-35. In

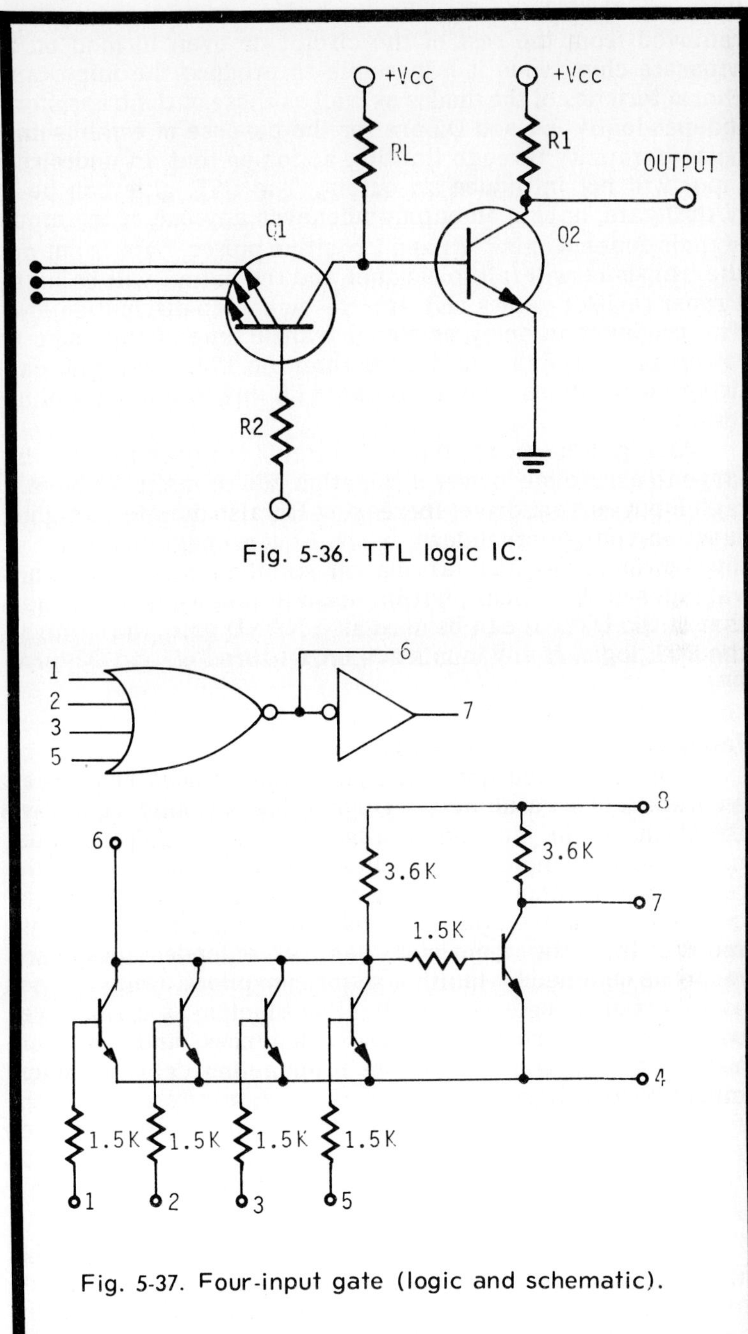

Fig. 5-36. TTL logic IC.

Fig. 5-37. Four-input gate (logic and schematic).

this case the diodes may be located in an isolated section removed from the rest of the circuit or even formed on a separate chip. Thus it is possible to produce the important characteristics of the diodes as well as those of the transistors independently. D1 and D2 are for the purpose of establishing noise immunity through limiting action so that an undesired input will not introduce an output. The DTL gate can be a NAND gate, having an output whenever any one of the input signals does not exixt. If 1 and 2 positive pulses, but 3 is not **on**, the transistor will fail to conduct and the output will be high. Proper resistor values (R1, R2, R3) will keep dissipation low. The propagation delay or signal transit time of this logic is lower than the RTL, but higher than the TTL. The DTL can drive more stages than a similar RTL due to lower coupling losses.

As Fig. 5-36 shows, the TTL circuit requires that RL be large to assure low power dissipation and to make Q2 have a high input current drive. Increasing RL also dictates a higher junction voltage breakdown in Q2. Low propagation delay at low power levels and fabrication simplicity are major advantages in this circuit, but its noise immunity is lower than that of the DTL. It can be used as a NAND gate, the same as the DTL logic. If any inputs are **on**, Q1 turns **off** and Q2 turns **on**.

Integrated Logic Components

The fabrication and the use of logic blocks require an awareness of several facts regarding the parasitic capacitive effects due to the junction isolating regions. Good design must minimize the number of such areas and thus reduce the overall capacitance by making more efficient use of the circuit area. Linear amplifiers, and switching circuits as well, require that the transistor has an extended frequency response obtained by limiting shunt capacitors. Delay time is so important in logic circuits that its reduction by use of fewer components, shorter transmission distances, and fast transistors is compulsory. Average propagation delay is determined by the formula:

$$\frac{\text{on delay} + \text{off delay}}{2}$$

In determining the operating speed of logic circuits, the importance of rise and fall times must not be underestimated. As a rule, these times can be controlled by using speed-up capacitors, plus diodes in the emitter leg. These concepts may be applied to any logic circuit in order to improve the rise and fall times, but transmission delay depends entirely on the structure and number of elements. Gates are available in many different configurations, and numerous modifications permit a wide variety of applications other than the specific function for which the unit was originally designed. As an example, the ever popular flip-flop can be made up by cross-connecting two gates. Or, a 4-input gate will work as a 3-input gate; simply ground one input. Fig. 5-37 shows a 4-input gate logic diagram and schematic.

An amplifier IC for digital computers is known as a buffer and is used to increase the fan-in or fan-out capability. By the power gain provided, the number of other units that can be connected in parallel to the input (fan-in) or output (fan-out) is increased accordingly.

There are many types of multivibrator (flip-flop) ICs, and by simply cross-connecting two gates, the multi can be made up quickly. The R-S is one example, and the J-K (Fig. 5-38) is similar except for the added clock input (normally designated T in the logic block). Flip-flops usually divide by two, but by proper connection, division by any number is possible, using a few more integrated circuits.

ECL

Emitter-coupled logic is shown in Fig. 5-39, a Motorola MC1019 one-bit full adder. This unit exhibits a typical

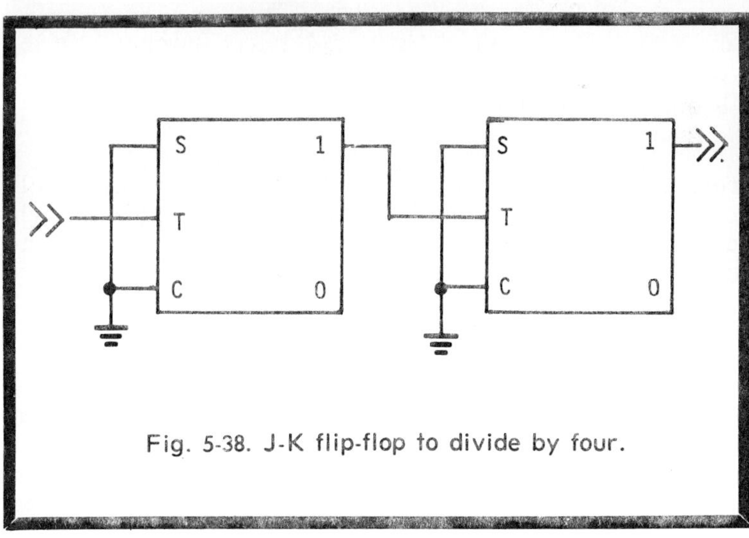

Fig. 5-38. J-K flip-flop to divide by four.

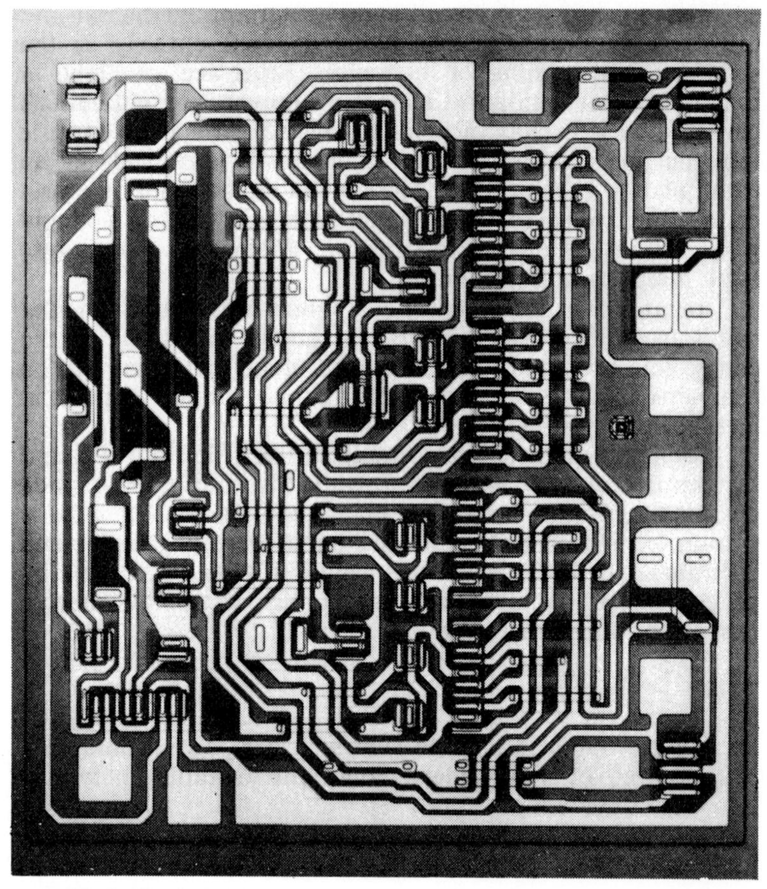

Fig. 5-39. MC1019 full adder IC. (Courtesy Motorola Semiconductor Products)

propagation delay of only 5.0 ns. Such an extremely small delay makes practical the implementation of ripple-through adders and multipliers for use in today's high-speed computers and instrumentation.

CENTRAL PROCESSOR

All the circuits discussed in this chapter can be found in a modern high-speed computer, such as the Burroughs B2500 and B3500 electronic data processing system. The central processor for this system is illustrated in Fig. 5-40. By means of Nixie tube and light displays, the operator is informed of the

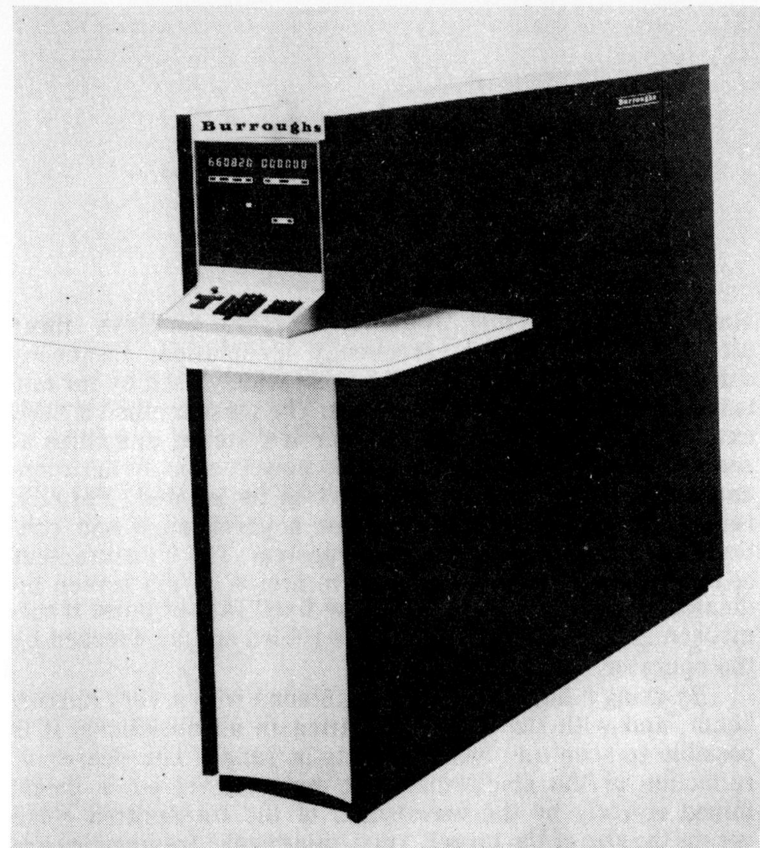

Fig. 5-40. Central processor for an electronic data processing system. (Courtesy Burroughs Corp.)

contents of the various registers and the settings of the logical indicators. A set of operating keys used in conjunction with the numeric keyboard allows the operator to read or alter internal memory or system control functions in any manner he desires. From here the system is programmed or instructed to perform the operations necessary to accomplish the work desired.

Chapter 6
Radar

Radio Detection and Ranging (Radar) employs three modes of transmission: frequency modulation, frequency shift, and pulse. The latter is the most widely used by far and is the form of greatest interest to us. The pulse method is used extensively by aircraft, ground search systems, and ships at sea. Pulse transmissions are high in power, short in duration, and beamed directly at the object to be located. Periodic repetition at a fixed rate provides a persistence and continuous display on the CRT of the receiver. The measurement of the time required for the echo to arrive on the screen indicates the range of the target. The fixed rate of pulse transmission allows time for the echo to return and be checked by the operator.

By using a highly directional antenna with a very narrow beam, and with the antenna rotating in all directions, it is possible to scan and locate objects in range. The degree of reflection of the electromagnetic waves involved is determined entirely by the wavelength of the transmitted wave versus the size of the target. Thus, microwave frequencies are used by all present-day radar systems.

Since the speed of light and radio waves is 984 feet per microsecond (for both the transmitted and the reflected wave), the actual distance to the target can be readily computed. The pulse repetition frequency is actually the number of pulses per second generated by the radar transmitter, and it is determined by the maximum range of the system, bearing in mind that sufficient time must be allowed for echoes to return before the next pulse is transmitted. Therefore, it is logical to assume that short pulses are needed for close targets, and longer (wider) pulses for distant objects. Pulse widths may vary from 0.5 to more than 2 microseconds. As the time between pulses is often as much as 1000 times the actual pulse width,

$$\frac{average\ power}{peak\ power} = \frac{pulse\ width}{PRT}$$

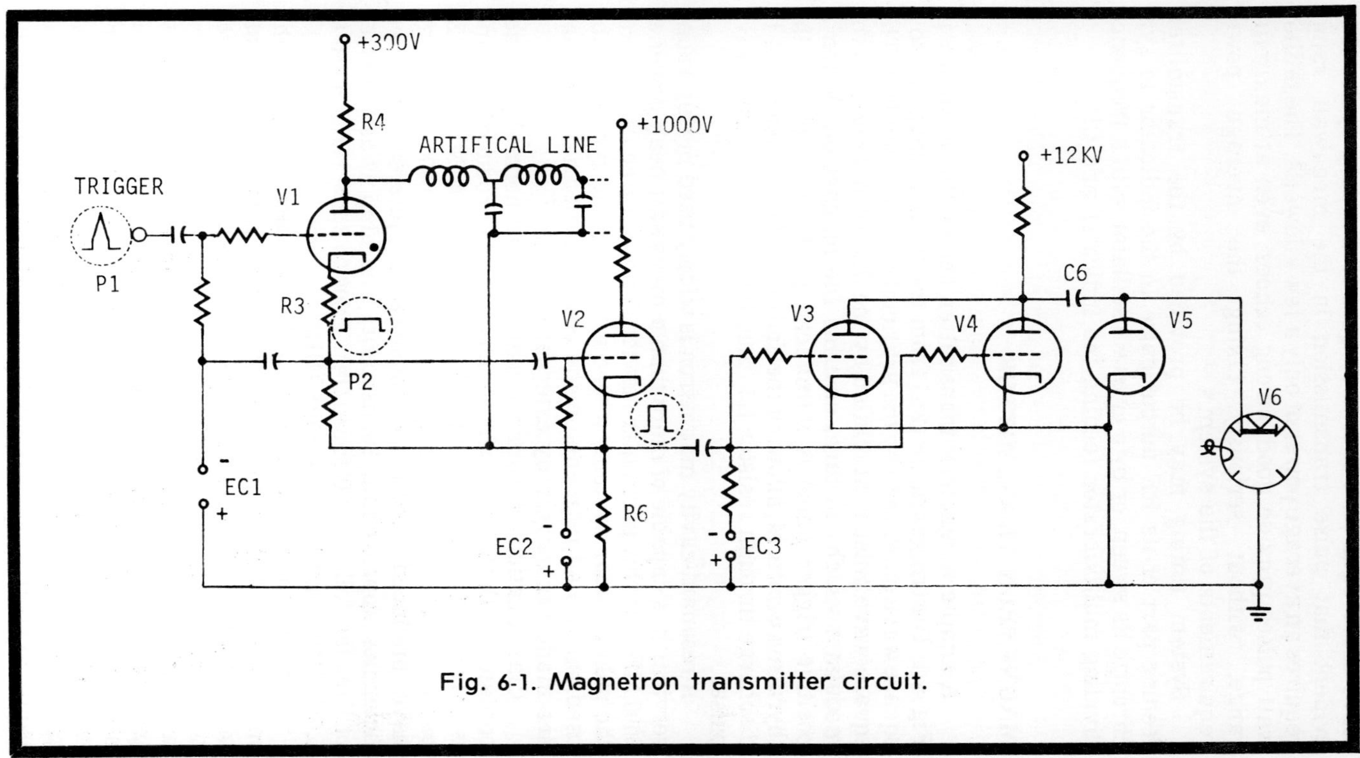

Fig. 6-1. Magnetron transmitter circuit.

Since these equations are also equal to the duty cycle, it is evident that pulse transmission in the megawatt range requires an average power of only a few kilowatts. The million watt pulses provide good, strong echoes, even at maximum range, without seriously taxing the average power requirements of the system.

System timing may be provided by the transmitter, feeding part of its RF output pulse to the indicator to synchronize its sweep, or by a crystal oscillator with a frequency dividing multivibrator feeding the indicator and CRT.

MAGNETRON TRANSMITTER

A simple magnetron transmitter schematic is shown in Fig. 6-1. The trigger pulse (P1) from the timer is shaped into a square wave pulse (P2) by the artificial transmission line. The square-wave pulse is amplified by the driver and controls the modulator which, in turn, controls the magnetron. When a positive trigger pulse is introduced to the grid of V1, the thyratron conducts, allowing the artificial transmission line to discharge through resistor R3, thus forming the square-wave pulse.

A resonant-cavity magnetron is widely used in ship radar service; it is capable of more than a megawatt peak power at 3000 MHz for this particular service. This is a fine example of the pulsed radar system, where a non-tunable type must be used, and all adjustments in the course of normal operation are made exclusively by external controls.

Classification of radar magnetrons is normally in accordance with the approximate frequencies produced. L-band is between 600 and 1200 MHz, S-band near 3000 MHz, and X-band about 10,000 MHz. Experiments in the 90,000-MHz area have produced results of considerable interest, with efficiencies approaching 90 percent. The ship radar service utilizes the following frequencies: 3,000 to 3,246 MHz, 5,460 to 5,650 MHz, and 9,320 to 9,500 MHz.

The magnetron plate voltage is the pulse voltage of the modulator, which means that the modulator must be capable of supplying a flat-topped pulse of constant amplitude, with the permissible variation not in excess of 5 percent. The magnetic field needed for magnetron operation is normally provided by a permanent magnet of a special high retentive alloy, since the magnetron must never be operated without the magnet. If it should be, excessive current will flow and probably damage the magnetron and modulator tubes. An overload relay is usually provided to prevent such damage.

158

Magnetron Operation

A magnetron consists of a rod-shaped cathode centered in a copper cylinder that forms the anode. As shown in Fig. 6-2, a series of cavities around the outside of the anode cylinder form the resonant tank circuits. A magnetron is operated in the field of a strong magnet, so electrons emitted from the cathode do not go straight to the anode. Instead, they are bent as a result of the strong magnetic field which they must pass through. The electrons whirl around the cathode in the same direction, which is determined by the direction of the magnetic field. As the whirling electrons pass the slots in the anode block, they couple energy into the cavities of the block, causing them to resonate. The cavities are of the proper volume to resonate at the desired operating frequency. Out-of-resonance electrons that pass the cavity slots extract energy from the cavity, causing current to speed up, and in so doing, the electrons are bent faster by the magnetic field; their path becomes curved to such an extent that they return to the cathode and never reach the anode. Because of their short path, as compared with the path of the working electrons, the out-of-phase electrons do not noticeably decrease the efficiency of the magnetron.

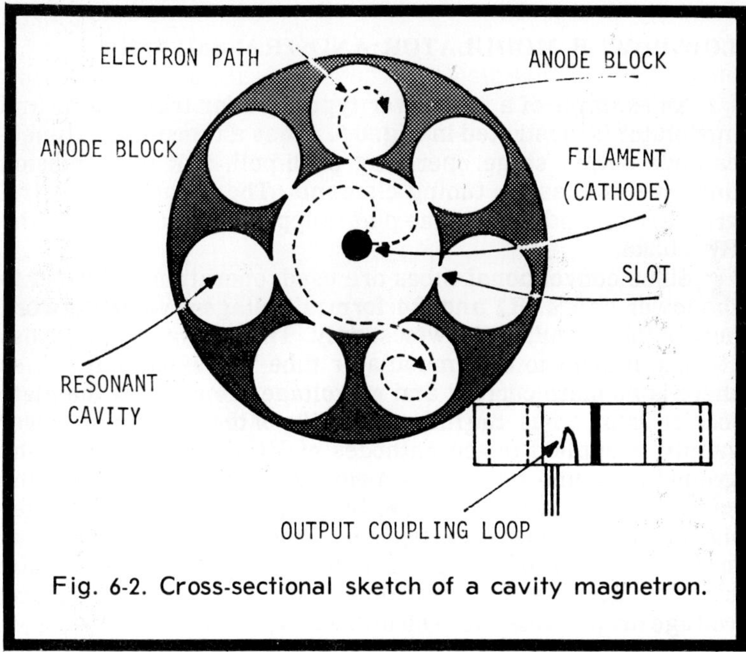

Fig. 6-2. Cross-sectional sketch of a cavity magnetron.

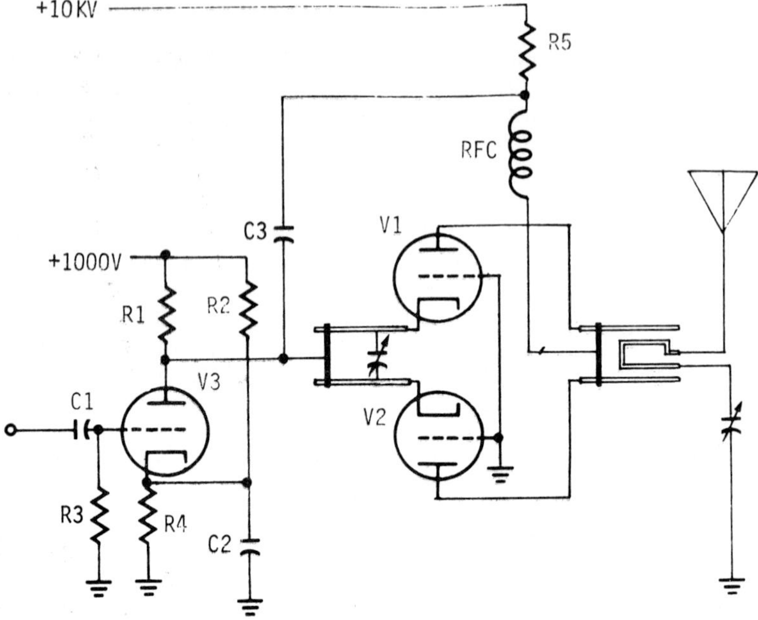

Fig. 6-3. Low-power modulator and transmitter circuit.

LOW-POWER MODULATOR AND TRANSMITTER

An example of a low-power type of radar transmitter and modulator is illustrated in Fig. 6-3. It has a tuned-plate, tuned-cathode output stage operating push-pull, and transmission lines are used as the tuning elements. The V1 and V2 control grids are grounded, with the plate supplied through R5 and the RF choke.

Since conventional tubes are used, operation is limited to the lower bands. R2 and R4 form a voltage-divider network across the modulator power supply. The voltage drop across R4 is sufficient to bias modulator tube V3 beyond cut off, so there is no plate current and no voltage drop across the plate load resistor. Thus, the full +1000 volts of the modulator power supply is applied to the cathodes of V1 and V2, making the grounded grids 1000 volts negative with respect to the cathodes. This bias blocks the tubes, making the transmitter inoperative during such periods. When a positive pulse of sufficient amplitude is applied to the grid of V3, the tube conducts and plate current flows through R1. The resulting voltage drop across the resistor reduces the grid voltage of V1 and V2 to a less negative value with respect to cathode. This unblocks the tube, allowing the transmitter to oscillate, and the resulting RF output is fed to the antenna by means of the coupling loop; C4 tunes the loop to resonance.

The oscillation continues for the pulse period of a few microseconds, and during this time the power to the antenna is high. As a result, the power input to the oscillator must be higher--actually much higher than the power supply can deliver. This additional power is supplied by a large storage capacitor (C3) which is charged during the long rest periods between pulses. The capacitor is charged through R5 which limits the charging rate so that the power supply will not be overloaded at any time.

C3 may also be used to determine the duration of the transmitted pulse. When C3 discharges during transmission, the voltage drops until it falls below the value required to sustain oscillation and the transmitter pulse stops abruptly. The size of the capacitor may be chosen to maintain continued oscillation for a period of one or two microseconds as required. Pulse duration may also be controlled by the modulator, and when such action is desired, the value of C3 must be large enough so that is will not discharge sufficiently to cut off oscillation.

Echo Box

An echo box is a test device that furnishes a standard reference signal for periodically checking radar systems. Acting as a microwave pulse generator, it is a high Q resonant cavity with provisions for coupling the signal in and out. Cavity length is adjustable by a plunger which in turn controls the frequency. When the transmitter is pulsed, the cavity resonator is energized and shock excited into oscillation. As a result of the high Q, oscillation continues for some time after the trailing edge of the transmitter pulse and returns a signal to the receiver, appearing as an artificial target on the indicator screen. The time required for the oscillations to drop from maximum to a level below the sensitivity of the receiver is called ring time. Ring time depends on power output from the transmitter, as well as sensitivity of the receiver. It is conceivable that losses could exist in the cavity resonator, but with the rugged tolerance requirements, the Q of the cavity is so high that any such losses can be completely ignored.

In the motor-tuned echo box, the cavity is periodically tuned through resonance for a brief period of time. The resulting target appears as a series of radial spokes on a plan position indicator. If the echo box is at fixed resonance, the target appears as an intensified circle in the middle of the PPI screen. The spoke length or circle diameter is measured when the system is known to be in good operating condition. With such a standard reference available, comparisons can be made with patterns produced at some later date. A decrease in spoke length is indicative of a decrease in performance in the system.

ANTENNA SYSTEM

The radar antenna is used not only for transmitting pulses, but also for the reception of the returning echoes. It is sharply directional and is mechanically rotated or "scanned." The principle of operation is the same as a parabolic antenna used at microwave frequencies, and in order to obtain the necessary directivity in an antenna of reasonable physical dimensions, microwave frequencies must be used for the operation of radar systems.

The waveguide, acting as a transmission line, is usually flared out to a horn at the end and is directed against a parabolic reflector which, in turn concentrates the RF energy in a narrow beam. Dimensions at the open end of the horn are critical, as the characteristic impedance must match the impedance of free space. Only when the horn is properly matched will the output be reflected away from the horn instead of back down the waveguide. The waveguide can also be terminated by a polystyrene window at the focal point of the parabolic reflector. By choosing proper window dimensions, the waveguide impedance can be matched to that of free space. Foreign material on the plastic window reduces efficiency, and if permitted to accumulate, the efficiency and accuracy of the system is appreciably impaired.

A radar antenna should have an unobstructed scanning area and a short waveguide run to the transmitter. Accessibility is also important for proper maintenance. Antenna design is based on the overall efficiency and the angular accuracy of the data to be obtained. Minimum waste of RF is the governing factor in obtaining maximum efficiency, and the transmission line loss between transmitter, receiver, and antenna must be held to a very low figure.

As the beam width is narrowed, the overall efficiency is improved. Location of targets in azimuth or elevation depends solely on the ability of the antenna to select the direction of the returned signal. Here the narrow beam is superior, and since side lobes or back radiation limit accuracy, the antenna system should be designed to hold these undesirable characteristics in the pattern to a minimum.

162

Transmission Lines

A coaxial transmission line is superior to an open-wire line because radiation losses are not encountered in the former as a result of the shielding, plus the fact that the outer conductor is at ground potential greatly simplifies installation. Centering and support of the center conductor is usually achieved by ceramic or polystyrene spacers, but some leakage still occurs at microwave frequencies. Special spacing reduces the loss somewhat, but stub supports consisting of a quarter-wave length of transmission line, shorted at the far end, produce much better results. The physical construction of the coaxial line does cause the use of stub matching to become quite complicated. Two stationary stubs of adjustable length are used in place of the single stub. Also, the spacing between conductors must be large enough to eliminate the possibility of arc-over during the peak power of the transmitted pulse. The transmission lines are normally filled with an inert gas and sealed, thus preventing moisture accumulation which increases losses and lowers the breakdown point. Even so, losses do become excessive at shorter wave-lengths.

Waveguide

With wavelengths below 10 cm, waveguides become a necessity. Aside from lower loss, waveguides do not need gas and are very easy to construct. Of course, it is possible for moisture to accumulate inside a waveguide, causing excessive losses, so a few precautionary measures are necessary to minimize this possibility. Long horizontal runs must be avoided. Gaskets are used where sections are joined with choke coupling flanges, and a drain is provided at the low point of the waveguide.

Choke coupling is much less expensive than direct contact flange joints which require precise machine work to give the waveguide a smooth inside surface at the joint. Perfect electrical contact and mechanical alignment at the joints, offering a smooth continuous inside surface entirely lacking in ridges, offsets, or rough spots, is the only way that the efficiency of the guide may be held within tolerable limits. In the case of choke coupling, the specifications are not nearly so critical. Although inferior to the direct-contact flange joint, the choke approach is still reasonably efficient.

Rectangular waveguide presents lower losses than the round type, but for a rotating antenna, the latter simplifies design techniques. A rotating joint bears some similarity to

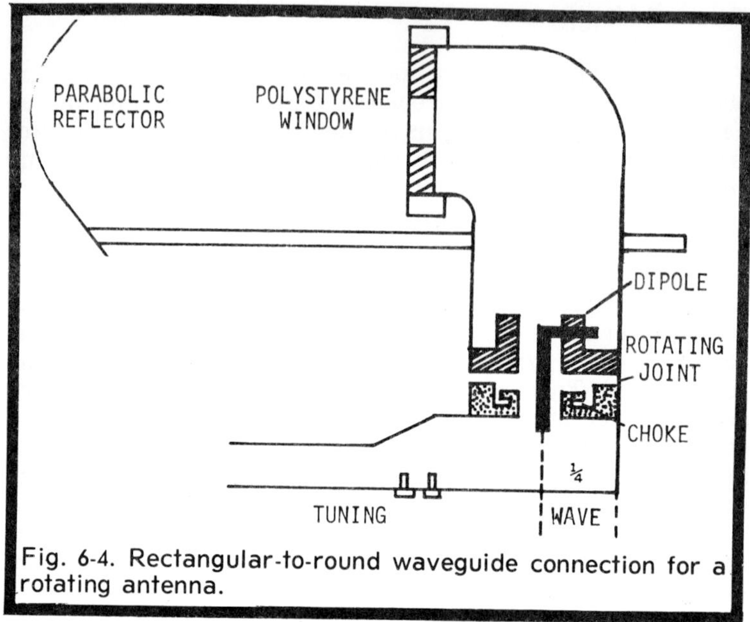

Fig. 6-4. Rectangular-to-round waveguide connection for a rotating antenna.

the regular choke coupling, except for the fixed gap, and one section is allowed to turn. This makes the choke simpler since the gap is fixed and the two flat flanges attached on either side of the break are less than one-tenth wavelength apart. The flanges, in turn, form the two sides of a quarter-wave open-ended line which presents a short circuit to the energy in the guide and thus confines it.

When coupling a round guide into a rotating joint, vertical and horizontal polarization result at different angular points. In order to acquire continuous vertical polarization, a short section of coaxial line is used in the rotating joint. A rectangular guide is coupled to a rotating joint by the use of a coaxial line as shown in Fig. 6-4. Two tuned plugs insure maximum transfer of energy by adjusting the impedance, and the inner conductor of the coaxial line extends into the rectangular waveguide. This serves as a voltage probe and enters the circular guide as a half-wave dipole. The outer conductor extends into the end of the circular guide about a quarter-wave, with the other half of the dipole fastened thereto. By breaking the outer conductor, we form the rotating joint simply by using a choke to bridge the gap. By bending the choke slot back on itself, the total length is a half-wave and the short at the end is reflected back as a short across the gap at this joint. Polarization results from the position of the dipole in

164

the circular guide, and since the dipole moves with the guide, polarization is not changed as rotation takes place.

RF directed from the guide to the antenna by the flared horn is radiated in a narrow beam. The gradual decrease in wavelength by the flared horn results in the match with the free space wavelength at the mouth, and perfect termination of the waveguide is attained. Termination by a polystyrene window of proper physical dimensions also results in an impedance match between the waveguide and free space.

Some radar antennas have the appearance of an inverted canoe which, with a width of approximately four feet, has a bearing resolution of two degrees. This enhances the radar's capability of separating targets in the same range and direction. The horizontal width of the radar beam determines the azimuth resolution, and 15 degrees in the vertical plane is usually sufficient to de-emphasize the rolling and pitching of a ship. Radar antennas or reflectors are frequently of cast aluminum or sheet-metal stampings with small openings (less than ¼ wave) to eliminate excessive wind resistance. Some manufacturers cover the entire assembly with a plastic radome.

RECEIVER SYSTEMS

Superheterodyne receivers are used in all radar systems because they provide exceptional sensitivity at a low noise figure with a wide bandpass. Mixing the received RF from the T-R box with the tuned local oscillator forms an intermediate frequency or difference signal in the 30 to 60 MHz range. As shown in the block diagram (Fig. 6-4), the IF amplifiers are followed by a second detector and IF filter, providing the video signal which is amplified in the following stages.

As the bandwidth-to-carrier frequency ratio is small, tuning is quite critical and AFC is required. A sample of the IF is fed to the automatic frequency control crystal stage, followed by an amplifier and discriminator. The DC output is passed through control circuits to the klystron local oscillator. Other methods can be used, and judging from the extensive progress made in television circuits, considerable improvements are available at the present time. The amplified video signal is applied to the indicator and determines beam intensity. Either electrostatic or electromagnetic deflection of the CRT may be used.

System timing circuits determine the PRR (pulse repetition rate), provide synchronization with the modulator and receiver, sweep and marker pulses for range information, and blanking pulses for the cathode ray tube. The time measure uses the instant of modulator firing as a base reference, and for this reason the synchronization between timer and modulator is of primary importance. The timer also controls the CRT, turning it on only during that portion of the pulse cycle when needed. The square wave generator performs other functions such as triggering the range marker generator and range sweep generator.

The range sweep generator produces a sawtooth wave to initiate the sweep and return it to its original position. The range marker generator provides video spikes at regular intervals during the range sweep. In both cases, the square-wave generator in the timer furnishes the pulse to trigger the switching action. If delayed sweep is needed, separate square-wave generators with delayed triggers are required.

Receiver Sensitivity

Overall efficiency of the radar system is dependent mainly on receiver sensitivity. Since the returning echo may have an amplitude of less than one microvolt at the receiver input, any slight improvement in sensitivity could increase the range of the system far more than a considerable increase in transmitter output. For good definition and accuracy, it is necessary for the signal to be well above the noise level after amplification.

Amplifier noise normally includes three general types: (a) thermal, (b) shot effect, and, (c) induced. All extend throughout the frequency spectrum and therefore, bandwidth has a direct bearing on the amount of noise in the receiver output. Unfortunately, the pulse is distorted as the bandwidth is reduced, although the noise level is lowered to a greater degree than the pulse amplitude. By using a wider bandwidth, the pulse is not distorted, but the noise and signal levels are frequently about equal. So, an adjustable bandwidth offers the best chance of a satisfactory compromise to meet a specific situation. On the plan position indicator, the noise appears as tiny points of light flashing at random, but due to the persistence of the screen, the targets are the brighter spots.

Because they have high noise and low gain, RF amplifiers are not practical at microwave radar frequencies. By locating the oscillator-mixer and first IF stage close to the antenna, excessive attenuation of the weak incoming signal is avoided. The main IF section provides the major portion of the receiver's gain and the input amplitude to the second detector must be high.

T-R and Anti-T-R Boxes

The T-R switch is also known as a duplexer and is basically an electronic switching device that makes it possible to use a single antenna for transmit and receive. The T-R electrically isolates the receiver from the transmitter during pulse transmission, so it will not be damaged by the high powered bursts of pulse energy. The anti-T-R box serves the purpose of electrical isolation of the transmitter from the receiver during its reception of the reflected pulses.

The spark gaps of the T-R and anti-T-R are exactly one-quarter wavelength from the waveguide. During pulse transmission, the high-voltage pulse arcs across both spark gaps. As soon as the arcing begins, the gap resistance becomes very low, and the input to the quarter-wave lines out to the gap "sees" a very high resistance (approaching infinity). Thus, most of the transmitted energy passes directly up the waveguide from transmitter to antenna. During reception, the spark gaps do not operate, so the resistance is high and the input sees zero impedance.

Receiver Blocking

Despite the large measure of protection provided by the T-R and anti-T-R devices, a high-level pulse from the transmitter can sometimes enter the receiver. This causes overdrive and may result in a temporary loss of gain in the IF. Thus reception of the reflected signals from nearby targets, which arrive shortly after the transmitted pulses end, is blocked. The blocking or lack of sensitivity of the receiver makes the radar system blind to signals from nearby targets. To minimize blocking effects, a rectangular disabling pulse is applied to the grid of one or more IF stages in order to produce cut off. Another method is use of an enabling pulse to turn the plate supply on or off at the correct times. The plate supply method is generally perferred.

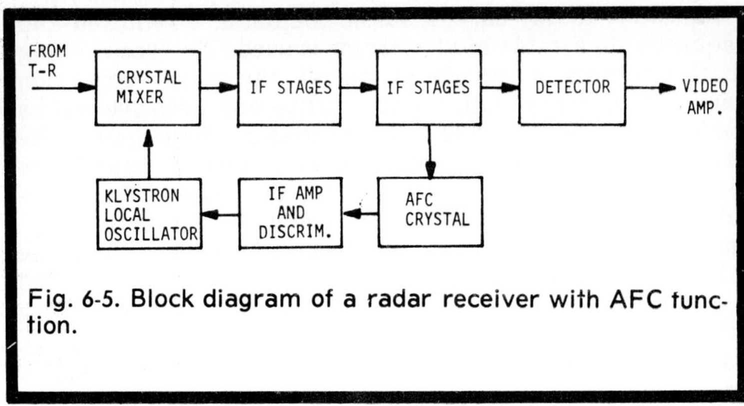

Fig. 6-5. Block diagram of a radar receiver with AFC function.

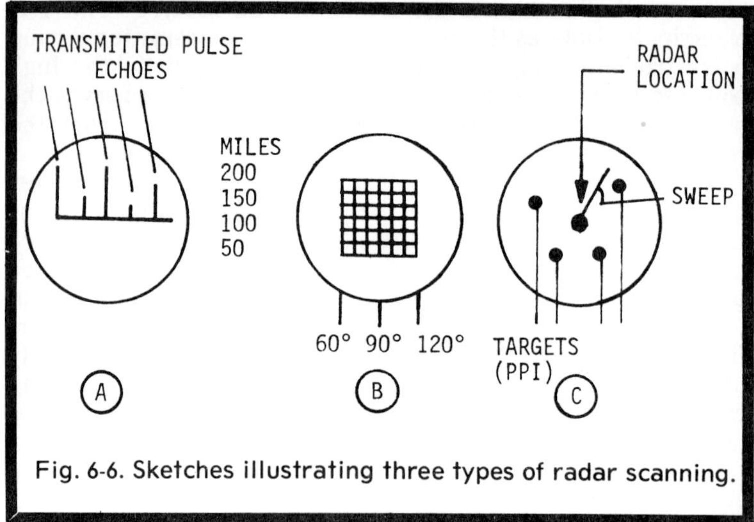

Fig. 6-6. Sketches illustrating three types of radar scanning.

INDICATOR

The indicator unit visually displays the time necessary for a transmitted pulse to reach its target and return. By means of a calibrated screen, time is interpreted in terms of yards or miles, and the distance or range to the target is indicated by a "blip." Basic components of a radar indicator are a CRT, sweep circuit, and a gate circuit. Various types of scanning are used as shown in Fig. 6-6.

RADAR TIMERS

The timer may be a separate component entirely, or it can be a part of other units such as the power supply, transmitter, or indicator. In any case, a stable oscillator is required and frequently an astable multivibrator is used. The free-running multi produces a square wave at the pulse repetition frequency of the system. Sharp positive and negative pulses are formed with the aid of a differentiator. The output of the differentiation circuit is applied to a diode clipper, the negative spike pulses removed, and the series of positive pulses sent to the modulator and indicator simultaneously. Thus, the modulator pulses the transmitter at the same time indicator sweep begins. The differentiator could be eliminated if a blocking oscillator were used; the blocking oscillator output would be passed directly to the clipper. The block diagram of a typical system in Fig. 6-7 shows the separate timer.

168

Several different voltages are required to operate a radar system. The transmitter places the major demand on the supply, since it uses a high voltage and heavy current but only for short intervals. A voltage-doubler circuit is adequate in most cases and filtering and requlation are not critical in the least. The CRT uses high voltage for the anode, but current requirements are small and a half-wave rectifier with an RC filter is sufficient. If a step-up transformer is used, a few turns (2 or 3) of heavy wire near either end of the secondary will provide the rectifier filament voltage. The second anode in the CRT is a colloidal graphite type conductor painted on the inside of the tube. The entire painted surface assumes the high-voltage potential applied to the second anode connection. This coating also forms an electrostatic screen to prevent external voltages from deflecting the electron beam. For the plate voltage of the receiver and indicator, a full-wave rectifier with choke input filtering for improved regulation is desirable. A bleeder resistor and voltage regulator are also quite useful here. As all high-voltage leads should be kept short, rectifiers are best distributed where needed, and the primary AC line carried to each point among the various units of the system.

DUTY CYCLE

The operating cycle is described as the fraction of the total time that RF energy is actually being radiated, or in other words, the ratio between average power and peak power.

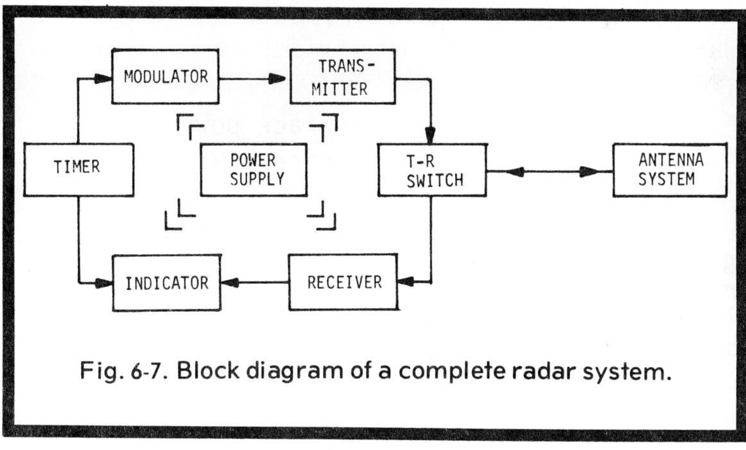

Fig. 6-7. Block diagram of a complete radar system.

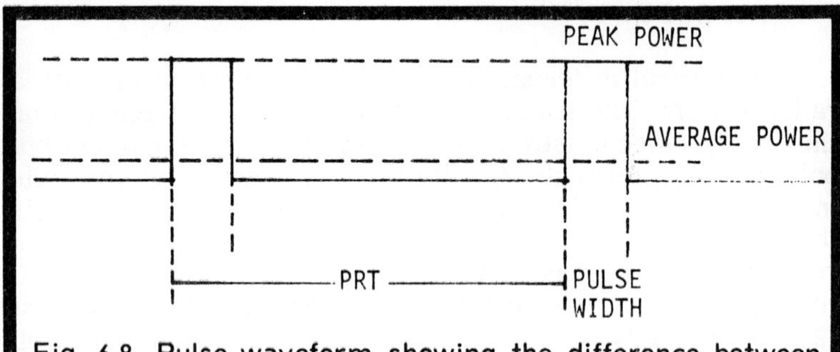

Fig. 6-8. Pulse waveform showing the difference between peak power and average power.

Average power dissipated over an extended period of time and peak power developed during pulse time are shown in Fig. 6-8. Pulse repetition time is equal to the reciprocal of the pulse repetition frequency. With other factors remaining constant, the longer the pulse width, the greater the average power; and the longer the pulse repetition time, the lower the average power. This is expressed in an equation:

$$\frac{\text{average power}}{\text{peak power}} = \frac{\text{pulse width}}{\text{pulse repetition time}}$$

The actual fraction of the total elapsed time that RF is being radiated is called the operating time; thus:

$$\text{duty cycle} = \frac{\text{pulse width}}{\text{pulse repetition time}}$$

or,

$$\text{duty cycle} = \frac{\text{average power}}{\text{peak power}}$$

As a result of the strong echo it produces at maximum ranges, high peak power is most desirable. The use of low average power permits smaller tubes and other components to be used, and this earmarks the advantage gained by a short duty cycle. The peak power possible is entirely dependent on the duty cycle.

170

Chapter 7
Telemetry

This complex branch of electronics is growing by leaps and bounds due to the increasing need for high-speed methods of communication. It provides many further uses for the pulse and switching circuitry covered in this book. The very foundation of our advancing technology is based on the information transmitted over amazing distances by radio telemetry. Although wires were originally used between sending and receiving stations, most systems have abandoned wire lines (except, some utility companies) in favor of electromagnetic waves.

Radio telemetry was first used for practical purposes in the early thirties to transmit weather information from balloons to the ground and later for the transmission of test data from aircraft. The systems were quite bulky, but as the need arose for smaller and smaller installations, designers kept pace and produced the required packaging, until today we have passed from vacuum tubes, to transistors, to integrated circuits (ICs), where we can have the equivalent of as many as 500 components in a chip measuring less than 5 X 7 one hundredths of an inch. So, as we continue to miniaturize, more problems are solved and progress continues at a phenomenal rate.

A basic telemetry system begins with transducers which convert physical or mechanical quantities into corresponding electrical quantities—a process called sampling. The sampling information modulates a transmitter which beams the combined data to a receiving station, where it is demodulated and recorded or used in some way. As we look into the various methods of telemetry modulation, many similarities are apparent between digital computer and telemetry switching circuits.

Telemetry reception is best achieved with circularly-polarized antennas and superheterodyne receivers with variable bandwidth. Maximum gain and selectivity, as well as

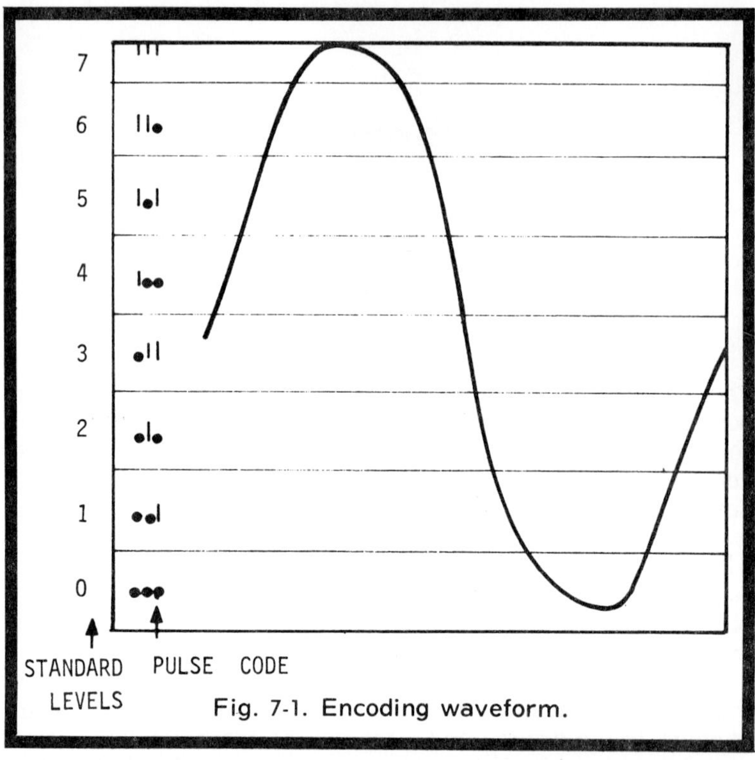

Fig. 7-1. Encoding waveform.

the best signal-to-noise ratio, are realized with dual-conversion receivers. Frequency stability is enhanced by the use of automatic frequency control, and bandspread or vernier tuning assists the operator in accurate operation.

Audio amplifiers are not used in telemetry, as meters are used to measure input and output signal levels. Test equipment, permitting a visual display of the subcarrier signals, allows the receiving station to check for noise, interference, hum, etc., and to analyze and correct problems encountered. By multiplexing and coding, we are able to further expand the information transmitted on one channel.

As the designer must be concerned with the possibility of noise interfering with the telemetry pulses, pulse code modulation (PCM) results in a worthwhile improvement in the signal-to-noise factor. By keeping the trailing edge of the pulse as steep as feasible, the introduction of crosstalk may be held at tolerable levels. Transducers useful in radio telemetry are usually of the analog type, although some digital units may be found in the field. The unbalanced trigger circuit discriminator is used in place of the ordinary PM detector, as

172

the latter is not sufficiently linear. Precaution should be taken at the receiver to insure careful alignment of the RF stages to minimize the possibility of error. In order to prevent distortion in the transmission of telemetry signals, the modulation applied to the subcarriers must be linear.

PULSE CODE MODULATION

In this system of modulation, information is conveyed by the arrangement of pulses in a series or train. The resolution or capacity of the system is limited only by the number of pulses per train. Fig. 7-1 shows this type of encoding graphically. The peak-to-peak range is split into several standard levels and correspond to a certain pulse arrangement. Thus the samples transmitted are not necessarily the actual amplitude but are the nearest standard. Some error, known as quantizing, is naturally introduced by this procedure, but the degree may be reduced to insignificant values by using a large number of standard amplitudes. Coded pulse trains are used to indicate the selected standard. Here each of eight standard amplitudes would be signified by combinations of pulses in a three-place code.

Pulses may be either present or absent in a binary numbering system. For instance, three pulses would indicate the number seven. The right-hand digit or pulse is (one), the next to the left is (two), and the final to the left is (four); the total is seven. Now suppose the line-up from left to right is: pulse, pulse, no pulse; we have 4 + 2 + 0 equals 6.

In a typical telemetering system, the beginning of each code group is marked by keying the transmitter at double power for four microseconds with a synchronizing pulse. Thus synchronized, the receiving decoder counts in binary by the presence or absense of a pulse at standard intervals. The next sync pulse differentiates one number from the next.

PCM offers a big advantage because once the signal has been converted, further error will not be likely as long as the signal-to-noise ratio of the receiver is good enough to prevent mistaking a large noise burst for a pulse. For this reason, receivers having low inherent noise are mandatory.

MULTIPLEX SYSTEM EVALUATION

Major multiplexing systems have advantages and disadvantages, and any selection would be largely decided from the standpoint of space requirements, power dissipation, and number of variables.

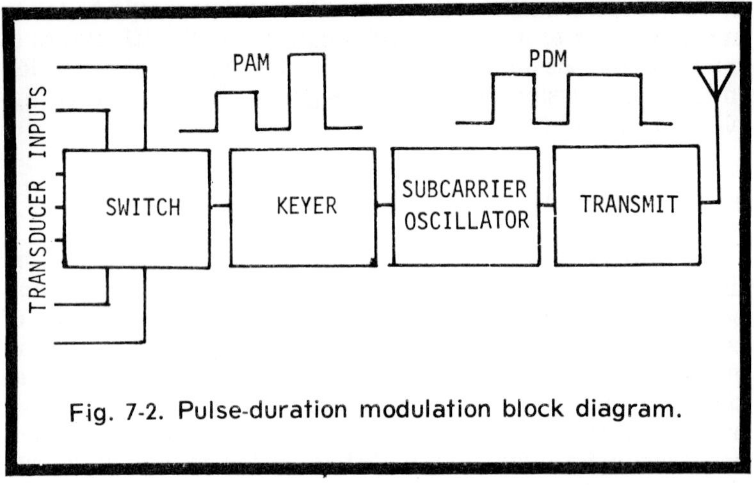

Fig. 7-2. Pulse-duration modulation block diagram.

Time-division multiplexing allows the output of several transducers to modulate a single carrier. When considering space requirements, time-division multiplex occupies less bandwidth than frequency-division multiplex for the same number of channels. But the frequency-division multiplex technique monitors each information channel continuously, and the high-frequency-handling capability is, therefore, superior to time-division multiplex. The distortion of data measured in time occurrences is generally much more than that measured in amplitude magnitudes in a time-division system.

Circumstances require the use of both major methods of multiplexing on occasion, and in some cases they are combined in the same telemetering system. Frequently, the best solution to the problem is the one that demands the least amount of precision in the smallest number of individual channels conveying information.

Multipath Effects

When signals are reflected from the surface of the earth or from objects on it, the desired signal may be received over many different paths, and as distances vary, the time of arrival varies in accordance. As may be assumed, multipath reception does cause considerable distortion in the majority of cases. In telemetering, such distortion may be reduced to minimum proportions by utilizing several receiving stations along the path of an airborne vehicle, and also by use of highly directional receiving antennas.

174

As an alternative to modulating the carrier by pulse amplitude, the intelligence can be impressed on the carrier by varying the duration of the pulse. This method is shown in Fig. 7-2, and is originated by the generation of an amplitude-modulated pulse. By applying the signal to the keyer circuit, it is converted into pulses of constant amplitude but with variable durations.

The basic keyer circuit is illustrated in Fig. 7-3. Its output is applied to the subcarrier oscillator. Pulses occurring at the commutation rate are produced by the trigger pulse generator and applied to the monostable multivibrator. If the commutator takes 60 samples per frame, with a frame rate of 15 per second the commutator rate would be 900 and the trigger pulse generator would deliver 900 output pulses per second. The PAM signal from the commutator is fed into the transformer primary of the keyer circuit and superimposed on the grid of V1. Each trigger pulse starts one cycle of monostable multivibrator action and is terminated after a time interval which varies linearly with the signal voltage induced into the grid circuit through T1.

The plate current of V2 is maximum when no signal is applied to the input terminals; therefore, the cathode of both tubes is positive enough to force V1 into cut-off. As the triggering pulse is applied to the grid of V1, action begins with the V1 plate current rising to maximum and the V2 grid is driven beyond cut-off. This condition prevails until the charge on the feedback capacitor leaks off and the negative voltage on the grid of V2 drops to where the tube conducts once more. The

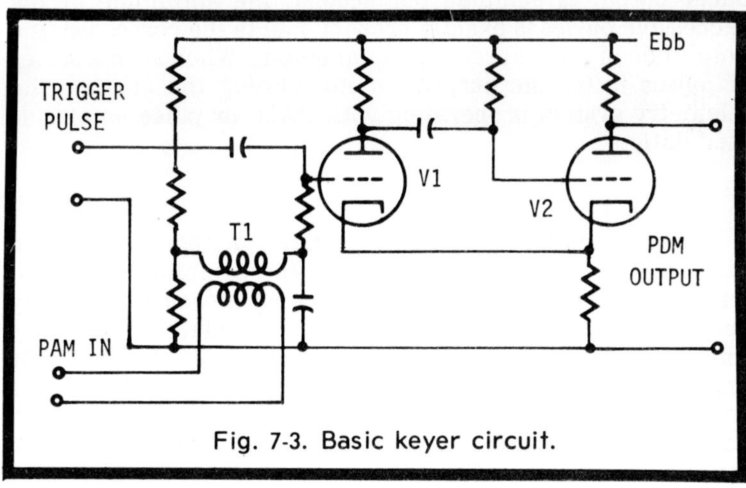

Fig. 7-3. Basic keyer circuit.

circuit then returns to its original state, and the length of time that V1 conducts and V2 is cut off is controlled by the signal from the commutator applied to the primary of T1. As the amplitude of the pulse from the PAM signal increases, the plate current of V1 increases, as does the grid bias to V2; the output pulse in terms of duration depends on the amplitude of the pulse from the PAM. Since the PAM signal pulses do occur at even time intervals, the leading edge of the keyer output pulse stays in a fixed position and the trailing edge of the pulse occurs at varying time intervals according to the input pulse amplitude.

The output pulse of the keyer is utilized to modulate the subcarrier oscillator, and its output modulates the transmitter. Should the number of transducer outputs be small enough that only one subcarrier channel is required, the subcarrier oscillator could be eliminated and the keyer used to modulate the transmitter directly.

Although the commutation rate of PDM does not change, the number of samples per frame and the frame rate may vary as long as the multiple of the two is 900. This figure may not vary. Synchronization in the PDM system is provided by the elimination of two successive pulses at regular intervals.

PULSE AMPLITUDE MODULATION

Since continuous monitoring of each transducer is not necessary in order to secure sufficient research data, satisfactory results may be obtained by sampling the output of the transducer a minimum of one time for each cycle of the commutator. Regardless of the number of times the transducer output is sampled per second, the amplitude of the successive pulses is exactly proportional to the corresponding amplitude of the transducer signal output. When such a series of pulses is for the purpose of modulating the carrier, the telemetry system is operating with PAM, or pulse amplitude modulation.

Sampling

When the pulse amplitude modulation method is utilized in radio telemetry, it is essential that the sampling rate is at least double the highest frequency being measured. Only by such a procedure is it possible to accumulate satisfactory information. Slowly varying outputs need be sampled only once per revolution of the commutator or in each full timing series of an electronic switch. Since noticeable changes could consume a lot of time, a low sampling rate would be adequate for conveying such information.

In cases where the transducer outputs display considerable variation, more frequent sampling is required in order to permit accurate detection of these appreciable changes.

A drawing representing a mechanical commutator used for generating the pulse-modulated output is shown in Fig. 7-4. Transducers 1, 2, and 3 produce voltages as shown, and these voltages are connected to the contacts on the commutator disc. Signals from the first transducer are connected to contacts 1, 4, 7, 10, and 13; the second transducer uses contacts 2, 5, 8, 11, and 14; and the third, 3, 6, 9, 12, and 15. The number of contacts required on the commutator disc depends on the number of transducers times the samplings per cycle. In addition, contacts are needed for synchronizing pulses and reference levels as well.

Each complete revolution of the contact wiper blade is a frame, and the number of frames per second is the frame rate. The total number of samples per revolution, when multiplied by the frame rate, provides the actual commutation rate. As the commutation rate increases, so does the number of channels required per second to carry that information, and the higher the frequency of the subcarrier needed. More channels per second demand a wider bandwidth, which in turn necessitates the use of higher subcarrier channels.

Bandpass Filters

The signal appearing at the output of the receiver is representative of the composite signal carried on the RF carrier as modulation from the transmitter. In order to isolate the individual subcarrier frequencies from the composite signal, bandpass filters with sharp cut-off characteristics are

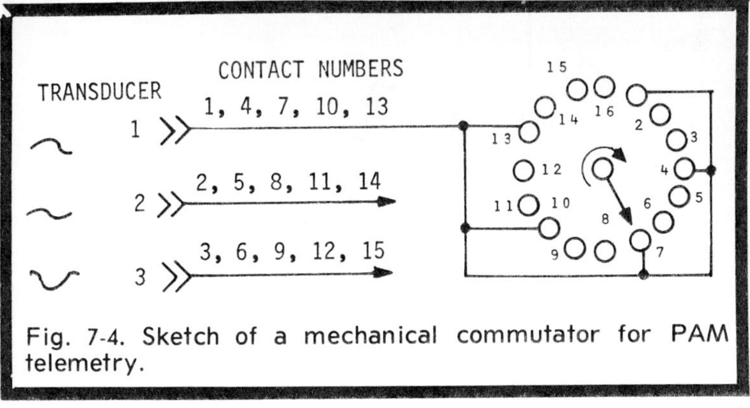

Fig. 7-4. Sketch of a mechanical commutator for PAM telemetry.

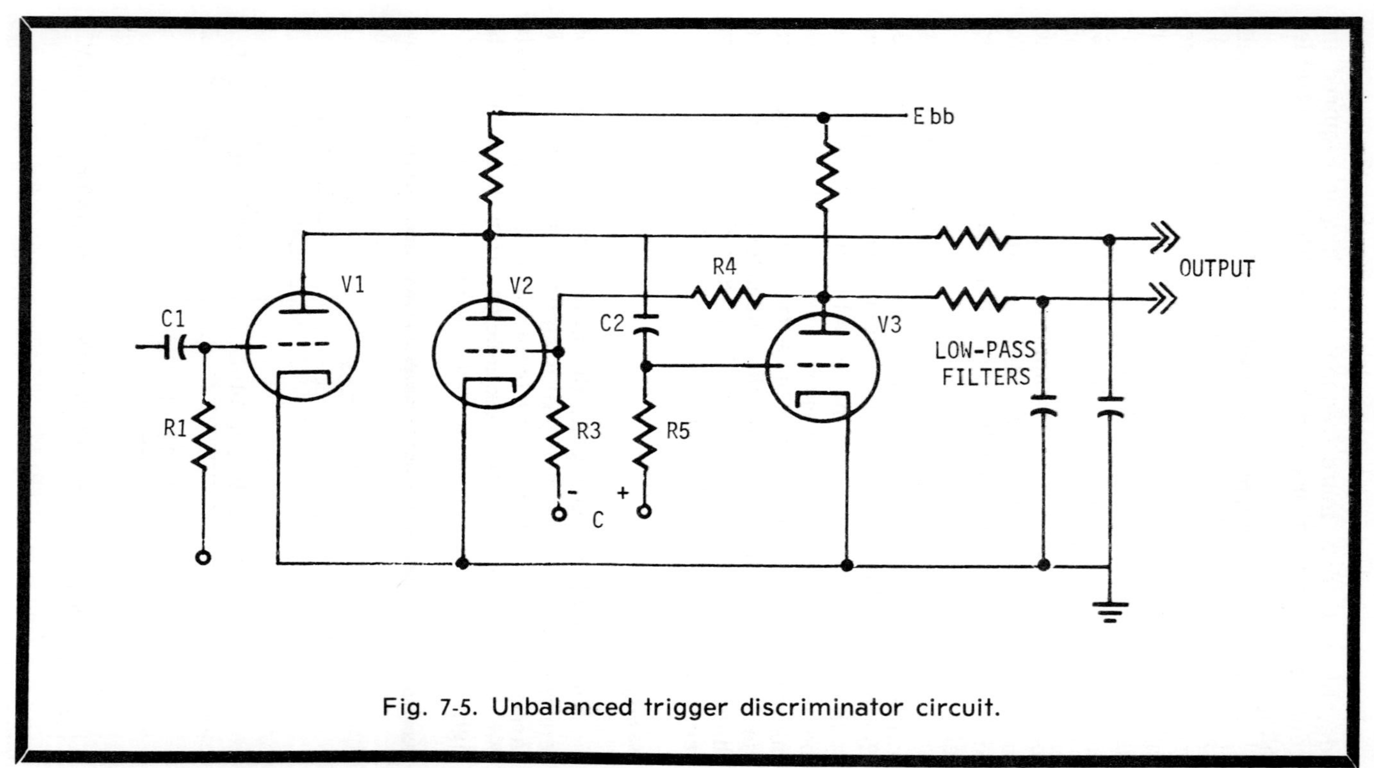

Fig. 7-5. Unbalanced trigger discriminator circuit.

required. A filter is used for each subcarrier channel, and it passes only the band of frequencies required for that particular channel. For example, using a subcarrier band with a center frequency of 1300 Hz, the bandpass requires a minimum range of 1202.5 Hz to 1397.5 Hz. These figures assume a maximum deviation of plus or minus 7.5 percent. The use of FM subcarrier oscillators is the more popular procedure, so the bandpass output is fed to the FM discriminator.

Demodulators

The bandpass filter output is converted by the demodulator into the original information as it was interpreted by the transducer. Ordinary FM detection circuits are not practical for radio telemetry purposes due to their lack of sufficient linear response. Since the usual circuits do not provide an output that is exactly a proportional replica of the frequency changes, the unbalanced trigger type discriminator, shown in Fig. 7-5, was developed. This discriminator presents an output of constant-length rectangular pulses, as it is triggered. A DC output having a zero voltage difference at center frequency in a push-pull circuit is made possible by this method.

The subcarrier signal from the limiter is applied to a differentiator and to the grid of the pulse generator (V1) which is cut off. V1 conducts only during positive pulses as a result. The trigger circuit discriminator, a monostable multivibrator using V2 and V3, is in its steady state when no input pulse is applied. With V2 cut off and V3 conducting in the steady state, the discriminator completes an operating cycle each time it is triggered by V1. During this triggering process, the monostable multi causes V2 to conduct and allows V3 to be cut off.

As the plate voltage of V3 increases to the value of the supply voltage, a positive bias is applied to the grid of V2, and forces the latter to conduct. V2 continues to conduct, with V3 cut off, until the feedback capacitor discharges sufficiently to permit V3 to start conducting again. As this takes place, the monostable multivibrator assumes its original stable state, completing the operating cycle. Thus the output is a near perfect linear function of the input from a frequency standpoint. The amplitude for a specific deviation in frequency depends on the relationship between the plate voltage cut-off value, which is normally equal to the supply voltage and the actual plate voltage while the tube is conducting; both values are determined by tube characteristics as well as circuit constants.

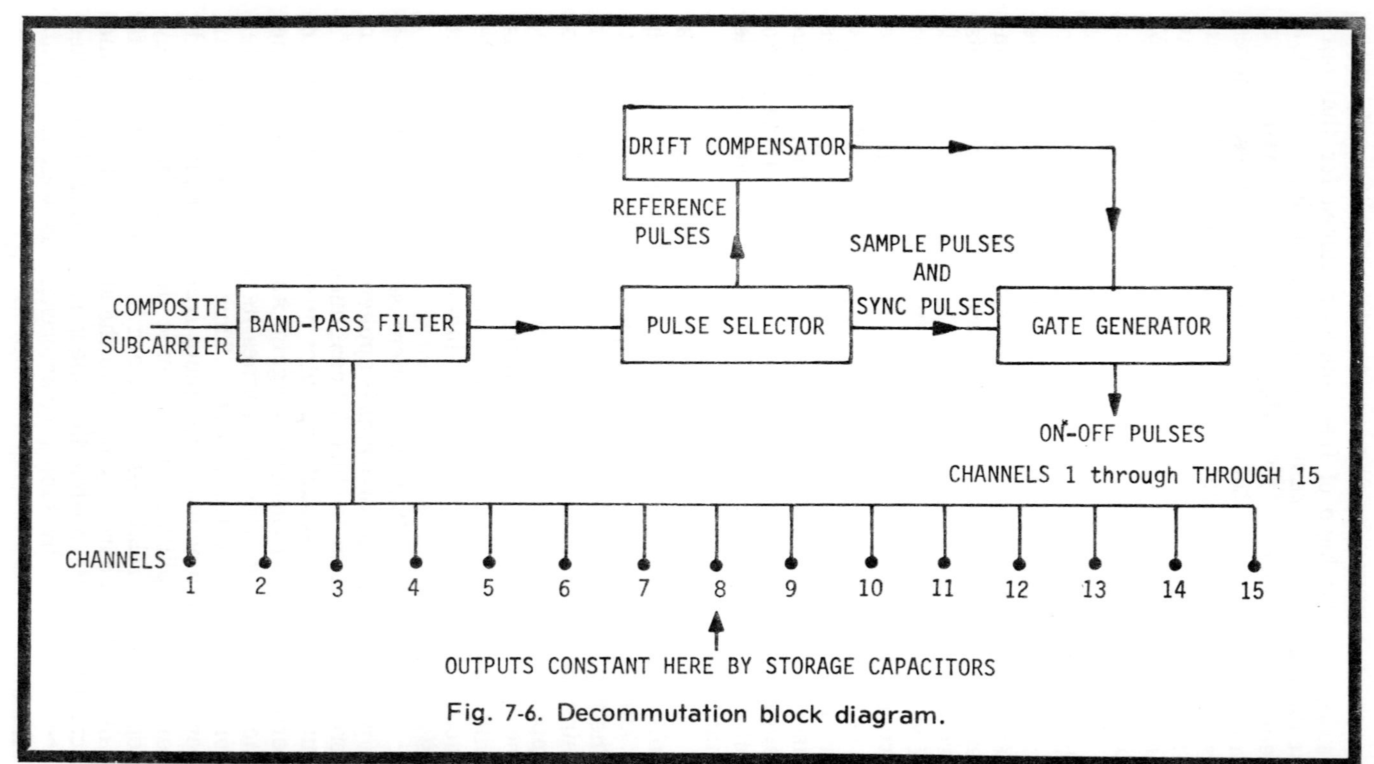

Fig. 7-6. Decommutation block diagram.

Since pentodes offer the important advantage of better linearity, their use is most desirable, along with precision (wirewound) resistors and other low tolerance components. The supply voltage is regulated, and the low-pass filters remove any undesired frequencies that could otherwise get into the output signal.

Because the discriminator output signal must duplicate the transducer output signal, it is quite feasible that some leakage could result from the many higher frequencies in use. The low-pass filters eliminate this likely problem, by allowing only the desired deviation frequencies to pass into the amplifiers. Such amplifiers are current types if visual reproduction through the use of a magnetic oscillograph recording device is desired.

Decommutation Procedures

Decommutation enables the recovery of the information in the PAM signal, and whichever method is selected it must be synchronized properly. The inertia factor in mechanical switching rules out this method, thereby necessitating the use of electronic switching. Many possibilities exist to meet the specific application at hand. Output signals from a decommutator may be offered as series outputs or parallel output. The former are presented in succession, while the latter are offered simultaneously.

In order to have the outputs available all at one time, the arrangement shown in Fig. 7-6 could be utilized. The composite subcarrier signal from the bandpass filter is offered to the pulse selector and 15 discriminators if a mechanical commutator were used. Generating a pulse for each of the consecutive information samples, plus synchronization and reference level, the pulse selector provides the output information to be applied to the gate generator. The reference pulses are applied to the drift compensator which determines if any change exists from the 100 percent and zero levels. The drift compensator offers an error-signal output to compensate for any drift in reference levels by applying a correction voltage to the gate generator.

The decommutator controls all output channels by switching the information to the proper discriminator in order. The discriminators convert the pulse amplitude modulated or pulse duration modulated signals to equal linear voltage changes, according to the system. Holding the output level for each channel, the storage circuit maintains a constant input to the readout device. After each frame, the stored

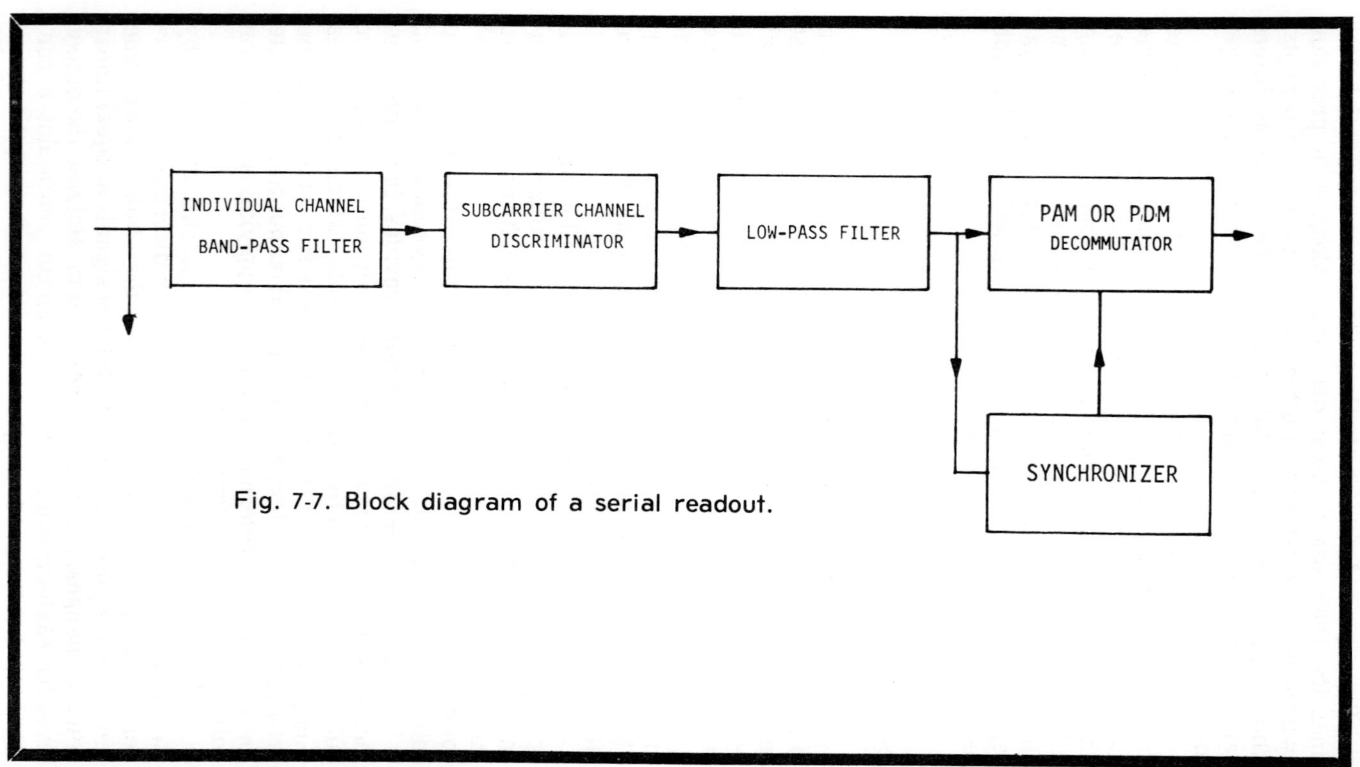

Fig. 7-7. Block diagram of a serial readout.

charge held in the capacitor is discharged, permitting any new information to reset the voltage level.

A block diagram of the decommutation arrangement for serial output is shown in Fig. 7-7. By introducing the composite signal to a bank of subcarrier bandpass filters, the respective subcarrier output from each bandpass filter is offered to the corresponding subcarrier discriminator, followed by its low-pass output filter. The recovered PAM or or PDM signals are finally applied to the decommutator or gate generator and the synchronizer. The decommutator generates an output voltage of corresponding value for each succeeding information pulse, and the synchronizer switches the gate generator on for each information pulse and off for synchronization, and also offers the sync pulse to control any desired external accessory.

MULTIPLEXING SIGNALS

For reasons of economical operation, parallel outputs are frequently applied to a single read-out unit, with an electronic switching arrangement for multiplexing. The output from each channel is successively applied to the read-out for a length of time determined by the switching speed and is prevented from reading a changing signal value by the use of capacitors in the multiplexer input lines. The ganged switch (Fig. 7-8) is closed long enough for the storage capacitors to charge to the exact level of the decommutated output signal in each instance. The ganged switch opens and the electronic switch rapidly discharges each capacitor in order for read-out. Following the last channel discharge, the ganged switch closes and repetitious operation continues.

RECORDING DEVICES

Numerous methods of keeping permanent records of telemetered intelligence for future study and analysis are available. By recording the discriminator output from the multiplexer, the exact quantities as originally measured by the transducer are reproduced at the convenience of the viewer and repeated as required.

D'Arsonval Galvanometer Oscillograph

The most widely used recording devices are the oscillographic type recorders utilizing galvanometers with mirrors attached to the coils. The coils are normally suspended by filaments and a permanent-type magnet sup-

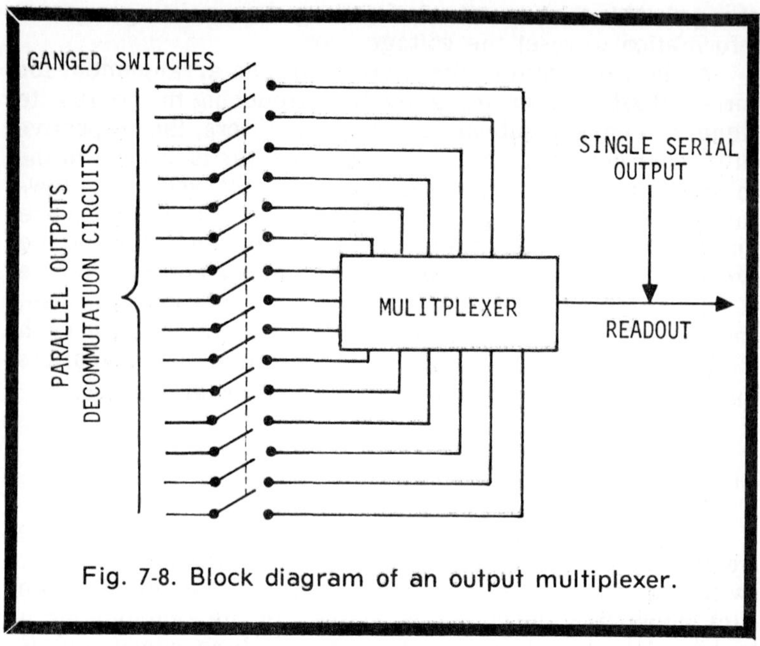

Fig. 7-8. Block diagram of an output multiplexer.

plies the magnetic field. Light from an incandescent lamp is reflected by the mirror onto a photo-sensitive film or paper mounted on a roll chart that is motor-driven at a uniform rate. Using a light baffle between the light and the mirror to narrow the beam for greater accuracy, the movement of the mirror varies the position of the light spot on the film or paper in accordance with the amplitude of the input signal. As the input signal is introduced to the galvanometer coil, varying positions relative to the magnetic field are assumed by the coil, and this movement is in direct relationship to the amplitude of the incoming signal. Needless to say, the activity of the galvanometer coil is transmitted by the filament wires suspending the mirror to position it in proper perspective to interpret the signal amplitude. The signal is recorded on the paper or film by means of the reflected light. Capable of handling 50 channels with 50 individual galvanometers, oscillographs of this type are commercially available. The frequency response of some galvanometer elements approaches several thousand Hz.

Video Recording

Recording the trace on a CRT screen by continuously moving photographic film provides a linear time scale. The

184

information pulses are reproduced as intensity modulation of the cathode ray beam, and channel timing is indicated by a series of pulses from a carefully synchronized generator. The pulses start and stop the sweep, and in this manner, any desired channel in order may be recorded. Two CRTs may be photographed on a single film; in fact, recording four channels on each tube is common, with proper timing indications included. This method of utilizing CRTs makes it possible to record the highest frequencies in the video bandwidth.

Pen Recording

Because considerable expense is entailed in the use of photographic recording, as well as unavoidable delays in the exposure and development of such film, a method of eliminating these problems is offered by the pen recorder. The use of a special type bearing and pen arrangement, attached to the moving coil of the D'Arsonval galvanometer, permits the coil movement to be transmitted to the pen, thus making a permanent record on the roll chart being drawn under it at a constant rate of speed. The use of several pen-galvanometer pairs produces side by side recordings of channels. By this procedure their comparison is greatly facilitated. The only point against this method is the limitation on the frequency of the signal due to the considerable inertia of the pen.

DATA REDUCTION METHODS

After telemetered information has been recorded, it is usually changed into a suitable form to be fed to a computer. The computer offers answers indicating the end result, or future work required. PCM information may be fed directly to the computer from the tape recorder, but in the case of PAM or PDM, the recordings must be converted to digital signals.

Chapter 8

Closed-Loop Radio
Remote Control

The pulse wave for radio remote control is quite simple and is transmitted on a frequency of 27.255 MHz, 53 MHz, 72 MHz, or 290 - 320 MHz, depending on the function and the type of license available. By F.C.C. regulations the length of transmission is limited to one second in each 30-second period. For this reason, timing circuits are incorporated in the transmitter so that a push of the transmit button nets a single one-second beep before cut off, no matter if the button is depressed constantly. After 30 seconds a second pulse of one-second duration may be transmitted, although this may not be necessary, as the initial pulse is normal enough to set the controlled device in operation. Unlicensed transmitting units are limited to 100 milliwatts of power, which, although low, is sufficient for reliable operation at short range without presenting an interference problem.

RADIO REMOTE CONTROL FOR HOBBY USE

Radio control transmitters operating on the 27.255-MHz, 53-MHz, and 72-MHz frequencies must be crystal controlled and require a Citizen's Band license, except on the 53-MHz frequency where an Amateur radio operator's license is necessary. Power output is normally about 1 watt, but some transistor transmitters operating in the 27.255-MHz channel provide excellent control up to 1 mile with only 250 milliwatts ($\frac{1}{4}$ watt). By using tone modulation of the RF carrier, multiple channels are provided for different control functions.

Model aircraft, boats, trains, and numerous other devices for hobby use may be accurately controlled by small radio transmitters and receivers. The transistor has proven a

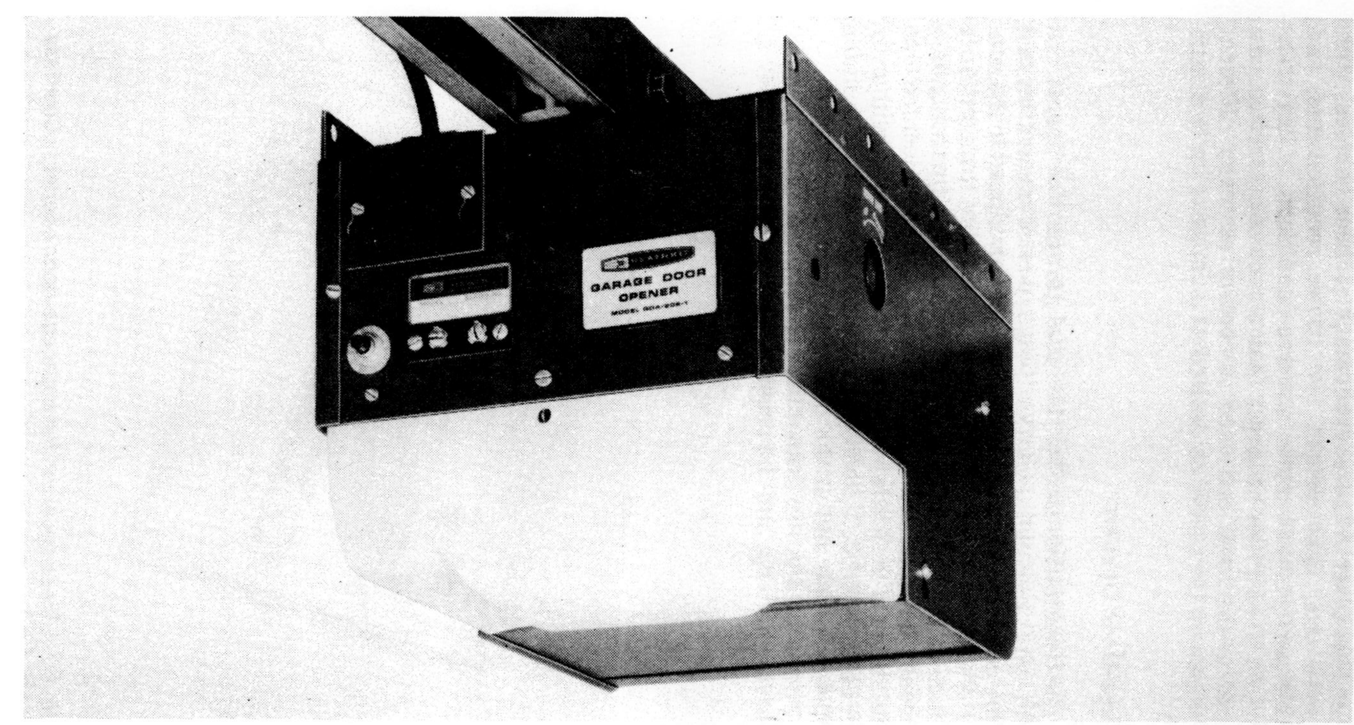

Fig. 8-1. Garage-door opener mechanism. (Courtesy Heath C0.)

tremendous asset to the advancement of these hobbies. With its small size, light weight, low power requirements, and reliable performance under adverse conditions, a tiny transistorized transmitter no larger than a pack of cigarettes can be held in the hand and used to control as many different operations on the model as desired at a range of up to a mile away.

Garage-Door Openers

The transmitter used for the usual garage door opener is a small, solid-state unit no larger than a pack of cigarettes, and readily carried in the glove compartment or clipped to the sun-visor of the car. Even so, the range of this tiny transmitter is as much as 200 feet with maximum reliability. Operating on a frequency between 290 and 320 MHz., it uses selectable tone modulation (12 to 23 kHz) to eliminate interference with or by other units. No F.C.C. license is required for operation in this frequency range and at the power provided.

Power for the tiny transmitter is supplied by the usual 9v transistor battery, and the receiver is powered by the 24v AC

Fig. 8-2. Linear receiver for a garage-door opener. (Courtesy Heath Co.)

line from the opener mechanism. Receiver range (sensitivity) is adjustable. A photo of a typical mechanism is shown in Fig. 8-1. The tiny receiver that picks up the transmitted pulse from the operator's car and triggers the mechanism is shown in Fig. 8-2.

Remote Light Switches

Remote control units may also be used to turn on outside lights from the car as it approaches the house. In the case of garage-door openers, most units will turn on garage or other lights as desired as an additional convenience. Other types of devices turn lights or appliances on at dusk and off at daybreak with photo-transistors or photodiodes.

A diagram of a positive acting (full or no load) light switch is shown in Fig. 8-3. It is capable of handling 2 amperes and operates from the 117v AC line. The switch, formed by Q1 and Q2, is of the regenerative type and offers a high impedance until the control voltage drop reaches a preset level. As the level is reached, the switch is triggered and conducts. The trigger point is determined by the resistance of the photocell and R4. When exposed to light, the resistance of the photocell decreases, causing an increase in Q2's base current. As this value increases, the transistor conducts and triggers the regenerative switch. The regenerative switch voltage is half-

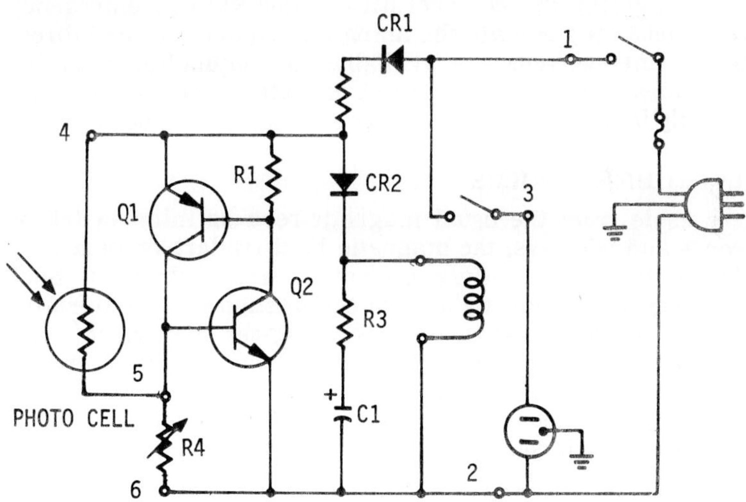

Fig. 8-3. Schematic of a positive-acting light switch.

wave AC rectified by CR1, and the switch may thus be triggered on each positive half-cycle.

If the photocell is not exposed to light, its resistance remains high and the regenerative switch cannot trigger. As a result, the relay across it remains actuated, releasing only when the regenerative switch conducts, shorting out the coil voltage and permitting the contacts that supply the external load to open.

Automatic Telephone Dialing and Answering

Pre-arranged telephone numbers may be dialed automatically, taped messages transmitted, or incoming calls answered and recorded by equipment triggered by radio remote control or through the AC wiring. Automatic dialing devices use tape or disc recorded dial pulses to make local, long-distance, or inter-office calls. Commercially available units normally accept from 30 to 40 "memory numbers," and operate from the AC power line. The number is dialed by a tone-modulated pulse which in turn triggers the dialing unit.

Telephone answering and recording machines operate with self-contained tape units, and "answer" the phone with a short pre-recorded instruction message. Upon a signal, the caller is invited to deliver his message to the recorder, with 45 to 60 seconds allowed for the call. The average cassette tape will accommodate about 50 messages. With this type unit, the tapes are easy to file for later use.

Triggering devices operating under various emergency conditions may actuate the automatic equipment, and direct phone line hookups are available in conjunction with the machines. It is also possible to bypass the phone and operate directly into the line when such an arrangement is desired.

INTRUDER ALARMS

Aside from the usual magnetic-reed or microswitch on doors and windows, the magnetic field (radar) or proximity (capacity) switch offers protection from intruders. Such switches are frequently used in conjunction with an infra-red beam, which triggers an answering-service phone. This protection is expensive, but reasonably positive, except against some professionals.

Although the intruder may defeat the door or window switch easily, it could be quite a task to out-smart a proximity or magnetic field type alarm. Many types of ultrasonic, radar, and magnetic field units and kits are available for the protection of offices, stores, shops, apartments, and other critical areas.

190

Proximity Switch Operation

The circuit shown in Fig. 8-4 is a basic proximity switch using a sensitive silicon controlled rectifier (SCR). Capacitors C1 and C2 form a voltage divider across the AC line, and resistor R1 is in series with C2, forming the critical reactance of the proximity sensor. Since C1 is small (10 pf), its reactance is large compared to C2. In fact, it is large enough to cause a voltage drop sufficient to trigger the neon lamp. This causes C1 to discharge through R2, providing a positive gate voltage to turn on the SCR when the line makes the anode positive. This causes current to flow through the relay or signaling device, which in the case of an intruder alarm latches in the activated position.

Although the SCR circuit resets as the cycle changes, the latching relay must be reset manually. A GE-5AH neon bulb is recommended even though a 10 percent loss could occur at the load with any neon bulb trigger. This is more than made up by their long life, reliability, and low price. The larger the area covered by the sensing loop the more sensitive the switch, because the capacity to ground introduced by an intruder's body is added to C2.

FIRE ALARMS

Home fire protection transmitters may be operated through the power line and will cover all circuits on the same

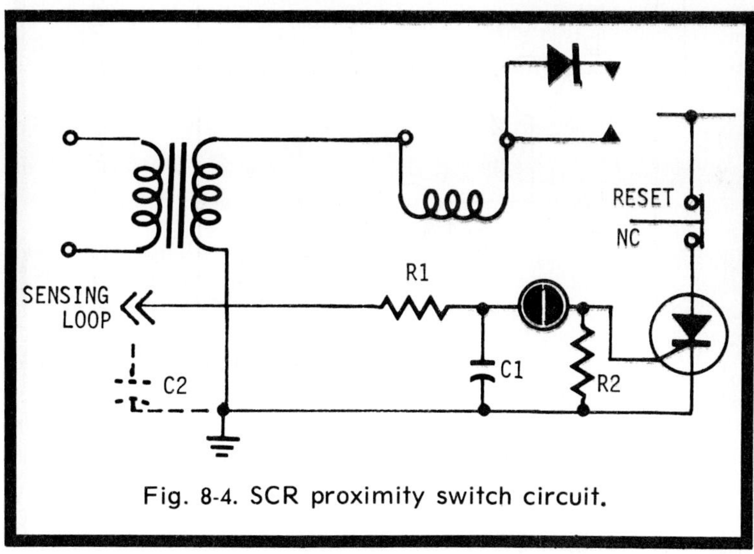

Fig. 8-4. SCR proximity switch circuit.

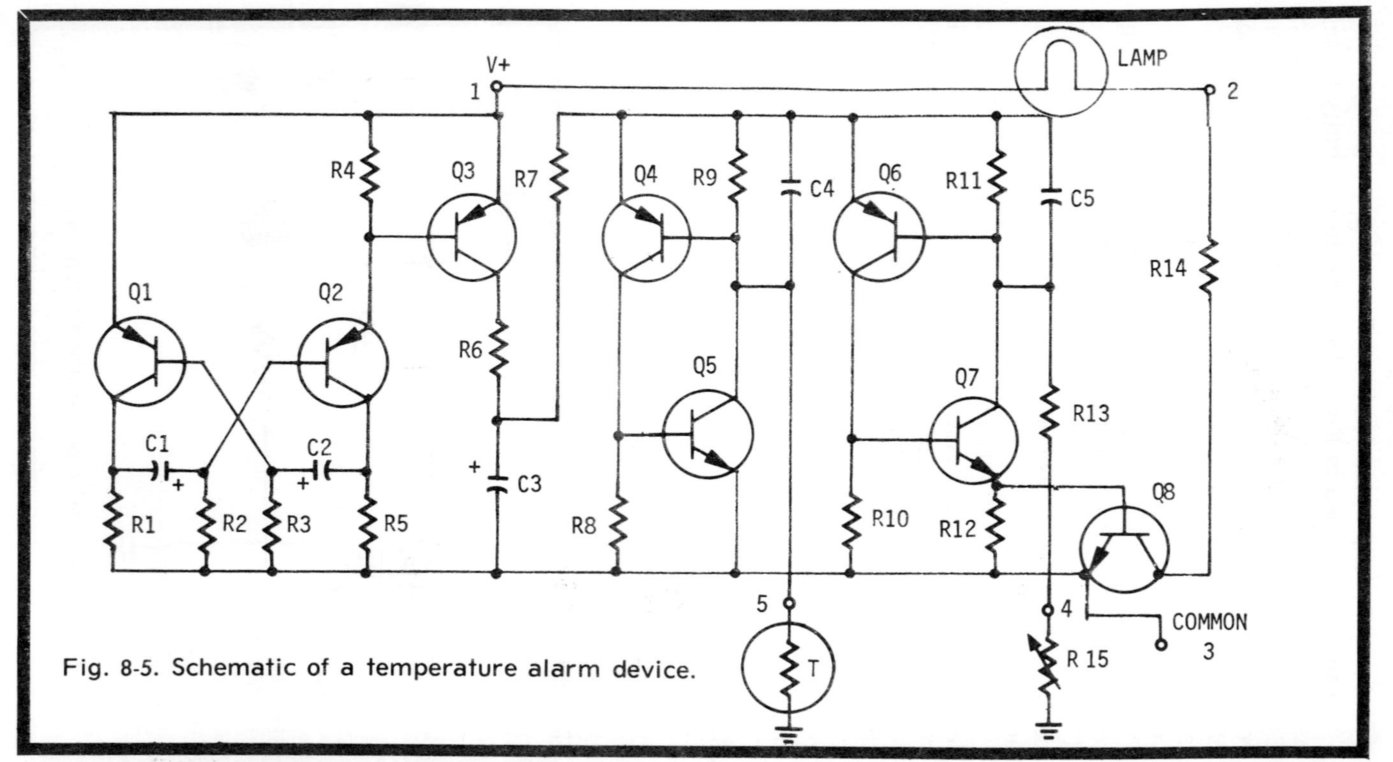

Fig. 8-5. Schematic of a temperature alarm device.

side of the transformer. Most units are solid state with printed-circuit design and the alarm utilizes a 2800-Hz transistor-type transducer. A smoke-heat detector needs no additional wires unless extra sensors are desired. Thermal-type sensors may be added to the alarm system where required.

The circuit shown in Fig. 8-5 is an extremely sensitive temperature-indicating device that may be used for many types of emergency conditions other than fires. The addition of a sound type signaling unit such as a bell, buzzer, or siren would make it suitable for a home protection unit. Transistors Q4, Q5, Q6, and Q7 form a voltage-dependent type of regenerative switch. The triggering level of the Q4-Q5 combination is controlled by the thermistor and the Q6-Q7 pair by sensitivity pot R15. Since the regenerative switches are in parallel, Q3 drives both from the intermittent voltage produced by the multivibrator (Q1 and Q2). This voltage assures sampling of the triggering voltages of the regenerative switches once per second, so the lowest switch conducts during the following second. Otherwise, the first switch to conduct would continue regardless of the status of the second switch.

While the thermistor resistance maintains a lower value than R13 + R15, the triggering voltage of regenerative switch Q4-Q5 is less than Q6-Q7, and the thermistor-controlled switch conducts to short out the other. As the thermistor temperature decreases, its resistance increases and eventually the triggering voltage of Q6-Q7 becomes lower than Q4-Q5. When this occurs, conduction shorts out the latter and current flows to the base of Q8, turning this transistor on to light the lamp.

The circuit is not affected by power supply variations since the triggering voltages of the regenerative switches are controlled entirely by the thermistor and potentiometer resistance. With the circuit arranged this way it can be used to warn motorists or farmers of falling temperatures. In order to indicate rising temperature the positions of the thermistor and R13-R15 must be reversed. The thermistor should be located in the temperature area for which warning is desired and sensitivity adjustment made at the borderline point where the warning device ceases to be actuated. An AC supply with emergency battery hookup (plus trickle-charging facilities to maintain the battery) are necessary when using the system as a fire-sensing alarm.

POWER LINE EMERGENCY LIGHT

Fig. 8-6 shows an emergency lamp circuit that is activated the instant power-line failure occurs. The lamp, in series with

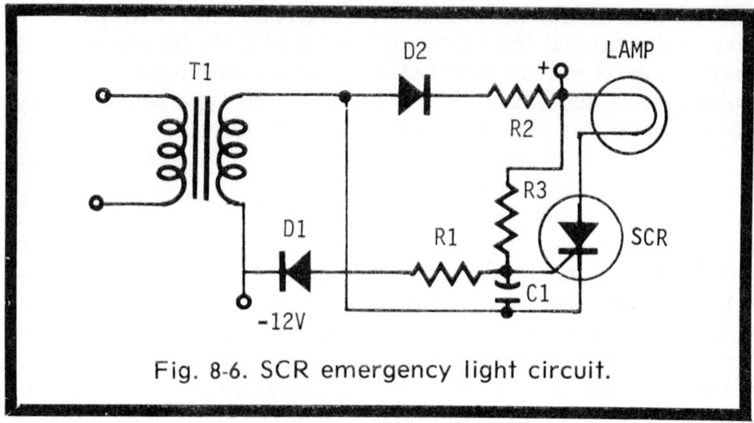

Fig. 8-6. SCR emergency light circuit.

the SCR and a battery, lights when the SCR turns on. As the SCR is normally off, the lamp is not lit. With AC applied to transformer T1, the top of the secondary is positive and D1 is conducting, causing C1 to charge to about 18 volts. Now C1's charge makes its lower plate negative, and as the SCR gate is negative the SCR is turned off. When the T1 secondary voltage reverses, C1 discharges somewhat through R2, R3, and D2, but this charge is renewed when the next half-cycle causes the bottom of T1 to be negative with respect to the top.

On failure of the power line, T1 no longer applies power to C1, so it discharges. The C1 discharge current flows from the negative (bottom) side of C1, through R2, R3, and D2, to the positive (top) side of capacitor C1. After discharging to a point where the D1 cathode is negative with respect to its anode, current flows from the negative battery through D1, R1, and R3 to the positive battery. This makes the R1-R3 junction positive to ground, having been negative while D1 was reverse-biased. The SCR is turned on by the positive gate, completing the battery-lamp circuit, and lighting the lamp.

When power is restored to the AC line, C1 is charged during the positive half cycle of T1, placing a reverse-bias on D1 which prevents it from conducting. This causes the SCR gate to be negative again (junction R1-R3), and as the SCR anode is negative during the next half-cycle, the SCR is turned off, extinguishing the lamp. The negative voltage at R1-R3 prevents the SCR from turning on until the AC power goes off.

BASE STATIONS (CLOSED-LOOP) REMOTE CONTROL

In order to control a base station from a remote point, the transmitter must be keyed and modulated from the remote point and the receiver output fed back to the control point. The

194

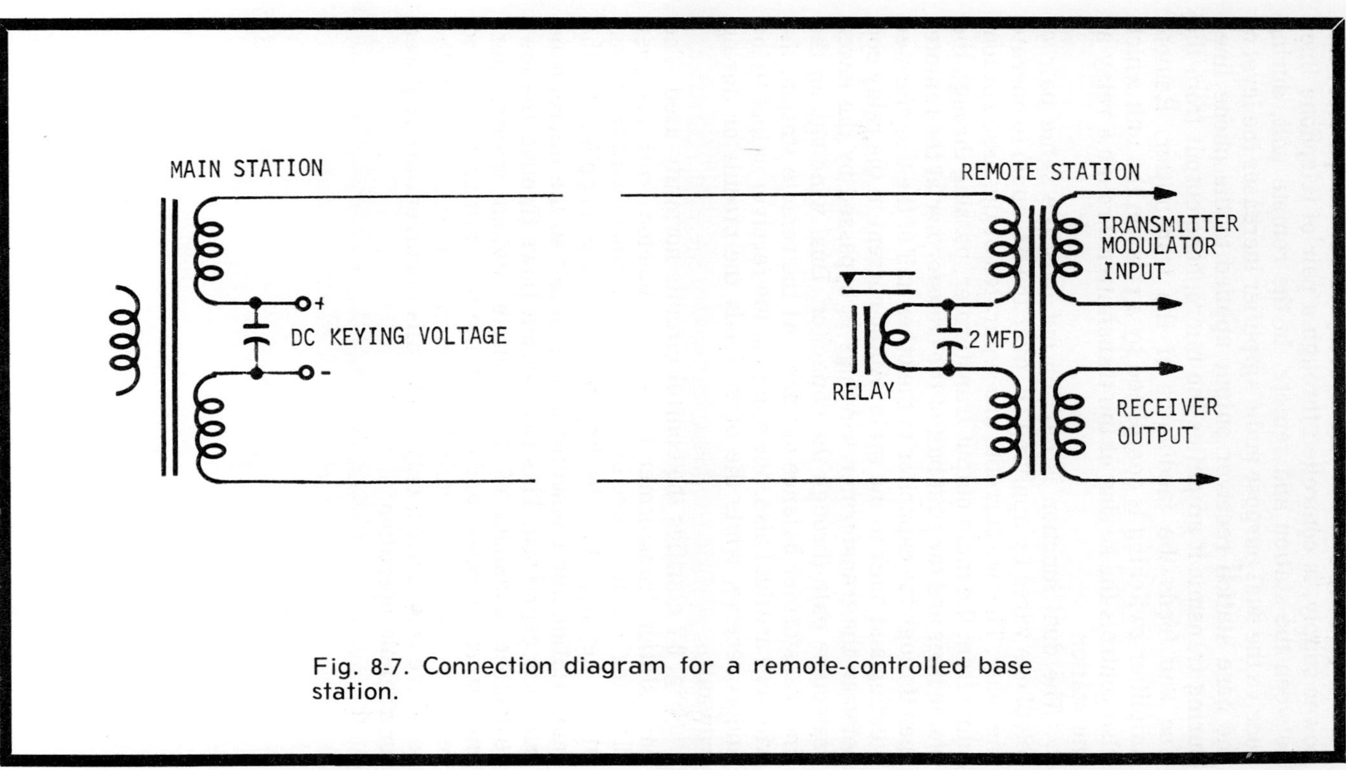

Fig. 8-7. Connection diagram for a remote-controlled base station.

remote package, consisting of relays, audio amplifiers, and power supply, is controlled through a pair of telephone lines between the station and remote. In the remote unit, during receive the dual-purpose audio amplifier increases the level of the base station receiver output applied to the phone line; during transmit it amplifies the microphone output from the line and feeds the modulator of the transmitter. Remote amplifier switching is completed by a relay in the unit which also controls the keying of the transmitter through a relay at the station.

The dual function is carried out over the same pair of telephone wires by applying DC over the AC audio to operate the relay. The two currents are separated at the base station, audio from the main output transformer, passing through the phone lines and the first half of the transformer at the remote, then through the capacitor to the other half of the transformer winding and back to the main. Audio current in the relay coil between the transformer windings is bypassed by the lower resistance path through the capacitor. Dual windings on the line transformer balance out hum. At the remote station, the windings are dual also, one applying the receiver output to the line in receive, while the other feeds the modulator during transmit.

Fig. 8-7 clarifies the control circuits normally used. The base station transformer remains connected, with receiver disabling accomplished in other circuits, along with the transmitter keying, etc. The keying current is supplied by the main station and connected to the circuit as the microphone button is depressed. This DC current flows through the low-resistance windings of the remote station transformer, energizing the relay, and causing the transmitter-keying and receiver-disabling circuits to be completed. As a result of the sensitive relays used, only a few mils of current are required for reliable operation.

196

Chapter 9
Automation

The thyratron, which was the first big improvement over the relay in power switching applications, has given way to the solid-state silicon-controlled rectifier in most equipment of recent design. The SCR has brought about considerable improvement in the capabilities of electronic systems in industry's automatic operations. The SCR is able to handle far greater power than transistors or vacuum tubes. However, although we can control the plate current of the tube by varying the grid voltage and the collector current of the transistor by its base bias, the gate of the SCR can turn the anode current full on but not off. In order to stop the anode current, it is necessary to interrupt it as least momentarily. Since it cannot be used to control the load current directly, the SCR must be used indirectly for this purpose.

The SCR has three terminals: anode, cathode, and gate or control. Four alternate layers of P-type and N-type silicon comprise the SCR, in PNPN order. The P anode is separated from the P gate by an N layer, ending with an N layer for the cathode terminal. Basically, it is a diode rectifier with conduction control. A positive pulse must be applied to the gate in order to turn it on, because it will remain in the off (nonconducting) state as long as the gate is negative or at zero potential. Once the rectifier is on, the gate loses all control until the anode voltage is interrupted or reversed. Then the gate again takes control and a positive pulse will turn the rectifier on again.

How do we interrupt the anode voltage? If AC is used for a supply source, there is no problem, as the SCR is reverse-biased for one half of each cycle. Conduction will not resume until the gate is positively triggered. Therefore, by maintaining positive bias on the gate, the load current is continuous.

SCR Characteristics

SCRs are available in voltages to 1500 volts or more with a current capability of many thousand amperes by parallel connection. Control gate voltage is normally 0, with positive firing voltage varying according to type and temperature of operation. A resistor is used across the gate input to hold it at zero potential and prevent accidental firing resulting from a positive build-up. The required positive pulse voltage for reliable triggering ranges from 1 to 3 volts, with the maximum allowable at 5 to 10 volts depending on the type. Since the impedance of the gate is low, the triggering source must be capable of supplying a pulse of sufficient current to fire the SCR. Firing currents may run as low as 10 or 15 ua in some of the more sensitive units and as high as 250 ma in heavy-duty units.

The low forward voltage drop of the SCR permits exceptional efficiency. Voltages greatly in excess of the power supply voltage are often encountered from inductance in the load, and damage to the SCR may result if this voltage surge exceeds the limits of the unit. In the form of a high-voltage spike, this may cause the SCR to switch on regardless of the gate bias state at the time. As described later, shunting diodes offer protection from surge triggering.

SCR Switch

The simplest SCR application is as a switch or mere replacement for a relay. This type of switching circuit is known as a static switch, since unlike the relay, the SCR has no moving parts. Static switching circuits fall into two groups: AC operation where line voltage reversal turns off the SCR, and those operating from a DC supply where additional circuitry is needed to turn off the SCR.

A static switch suitable for AC-operated loads is shown in Fig. 9-1. Notice that the SCRs are connected in inverse parallel, the anode of each to the cathode of the other. Each SCR conducts on alternate half cycles to supply full-wave AC power to the load. When the switch of the control device is open, each gate is held at its cathode potential (zero voltage on gate). The low resistance of R1 and R2, which connect between gate and cathode, assures zero gate voltage; therefore, the load is turned off. When the switch is closed and the right side of the AC supply voltage is positive, D1 conducts and supplies current to the gate of SCR2, turning it on; current then flows through the load. Resistor R3 is chosen to limit gate current to a suitable value, as the voltage drop across R1 may negative-bias the gate of SCR1 beyond its rated voltage value and thus damage the unit. If diode D1 were not in the circuit, the SCR2 firing current would flow through R1.

On the other half cycle the operation is similar, SCR1 now being the conducting unit. The positive voltage needed to fire SCR1 is supplied through D2 and R3. The control device itself can be either electrical or mechanical. Light-dependent resistors, magnetic reed switches, magnetic cores, thermostats, pressure switches, and timer switches are all suitable control units.

TRIAC SWITCH

The fact that the SCR is a rectifier is often a disadvantage. For example, in Fig. 9-1 it was necessary to use two SCRs in order to furnish power on both halves of the AC cycle. The symmetrical switch (also called a tri-lead AC switch or bi-directional AC switch) differs from the SCR in that it conducts in both directions. Tri-lead AC switches are widely known by the trade name Triac. Fig. 9-2 is a construction diagram and the symbols used to represent a symmetrical switch.

The many P-type and N-type areas make the structure approximately equivalent to that of a PNPN rectifier in parallel with an NPNP rectifier. Each one conducts on alternate half-cycles to make the device bidirectional. Terminal T1 corresponds to the cathode of an SCR in the sense that it us used as the reference terminal for gate and T2 voltages. Either a positive or negative pulse (with respect to Ti) applied to the gate will fire the Triac.

One practical triggering method is to use a positive gate voltage (with respect to T1) when T2 is positive, and with a negative gate voltage when T2 is negative. This triggering

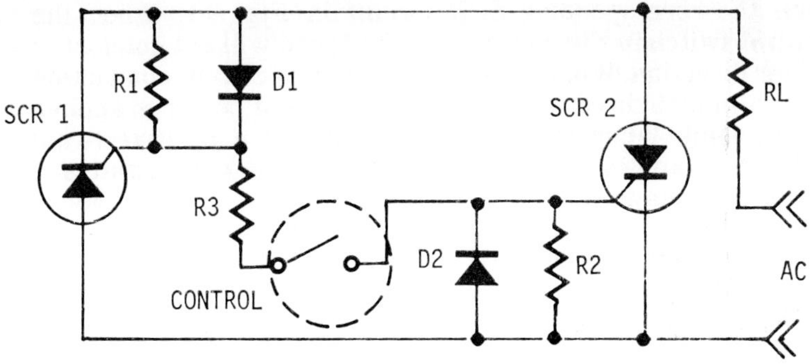

Fig. 9-1. SCR switch circuit.

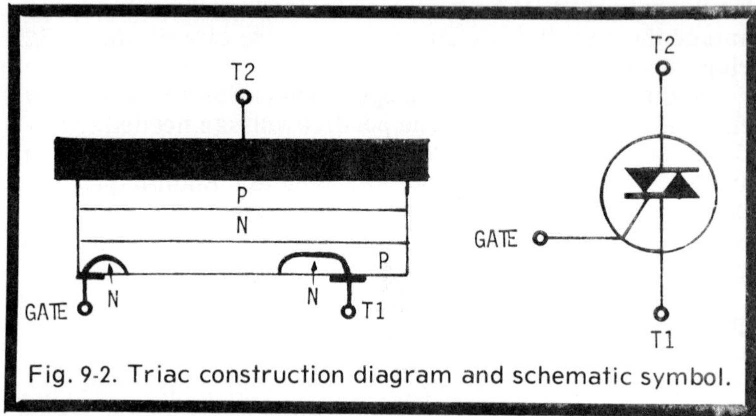

Fig. 9-2. Triac construction diagram and schematic symbol.

method is convenient when the triggering voltage is taken from the AC power supply. The only other commonly used triggering method is to use a negative gate-firing voltage for both half cycles. This method is appropriate when the firing pulse is supplied by a pulse generator. The pulse generator must supply negative pulses (with respect to terminal T1).

Triggering the Triac

Triggering a Triac with positive pulses is not generally suitable because triggering sensitivity is poor for a positive gate voltage when terminal T2 is negative. As with the SCR, the gate cannot be used to cut off a symmetrical switch. However, the switch is cut off twice each cycle when the voltage drops to zero between alternations. Thus the gate must be triggered each alternation to maintain conduction

Fig. 9-3 is a basic circuit used to control a load by means of a symmetrical switch. Notice how much simpler this circuit is than the corresponding SCR circuit in Fig. 9-1. When the control switch in Fig. 9-3 is open, the Triac will not conduct in either direction. When the control switch is closed, conduction occurs in both directions and the full cycle of power is applied to the load. When terminal T2 is positive a positive firing voltage is applied to the gate, and when T2 is negative a negative voltage is applied to the gate.

LIGHT-ACTIVATED SCRs

There are silicon controlled rectifiers that can be turned on by a beam of light striking the unit. One type of photosensitive SCR has no gate lead; the light beam serves the same purpose as the gate in an ordinary SCR.

Another type has an external gate so the SCR can be controlled either by a light beam or by the gate in the usual manner. In any case, the light beam, like the gate, only turns the SCR on; it cannot be used to turn the SCR off. Symbols for light-activated SCRs appear in Fig. 9-4.

UNIJUNCTION TRANSISTOR FOR SCR CONTROL

So far we have discussed circuits that turn power off and on. SCRs also can be used in more important ways, such as varying the power supplied to a load. For this purpose, pulses are used to trigger the gate. The unijunction transistor is widely used for forming the SCR triggering pulse. Later we show how the pulse-controlled SCR is able to vary load power.

The unijunction transistor has three terminals, but differs from ordinary transistors in that it has only a single PN junction. Comparison is really unimportant as its operation in the circuit is quite different from a transistor. The unijunction transistor solely as a switch, having no amplifying ability. It is used in sawtooth and pulse oscillators, timing circuits, voltage- and current-sensing circuits, and bistable multivibrators. It is particularly useful for triggering SCRs.

Basically, the unijunction (abbreviated UJT) consists of a bar or cube of lightly-doped N-type silicon to which a single P-

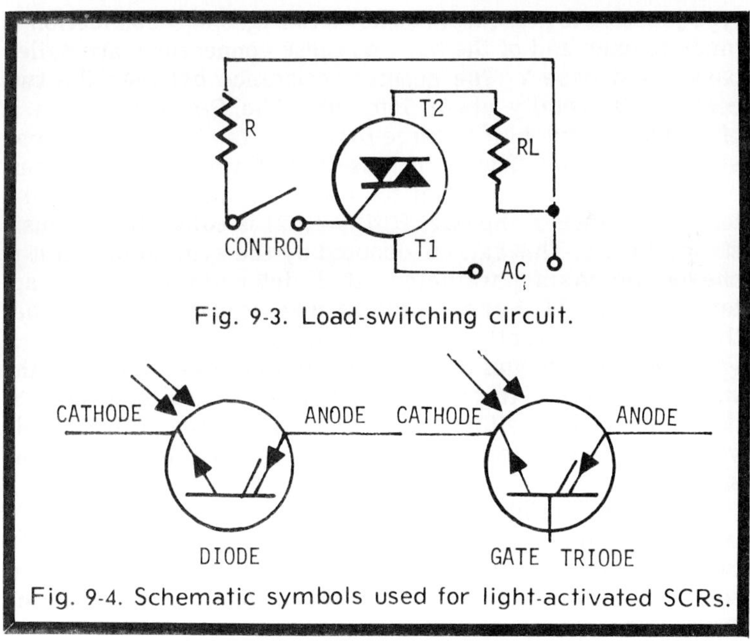

Fig. 9-3. Load-switching circuit.

Fig. 9-4. Schematic symbols used for light-activated SCRs.

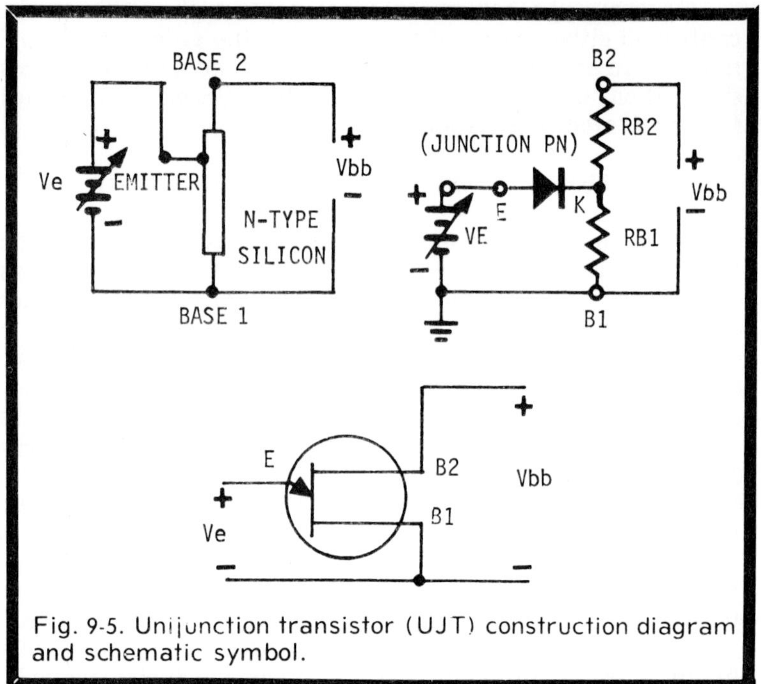

Fig. 9-5. Unijunction transistor (UJT) construction diagram and schematic symbol.

type junction is attached, as shown in Fig. 9-5. A connection is made to each end of the bar and these connections are called base 1 and base 2. The normal resistance between the two contacts is usually about 5 to 10K. The P-type junction is placed about one-half to three-fourths of the length of the bar away from base 1. The bar is thus divided into two resistances.

The ratio of the emitter to base 1 resistance (Rb1) to the total resistance of the bar (Rb1 + Rb2) is called the intrinsic standoff ratio. This ratio is denoted by the symbol n, and it is one of the main parameters that determines the characteristics of a UJT. For example, an n value of 0.5 indicates that the junction is exactly halfway between base 1 and base 2. A value of 0.75 tells you that the junction is three-fourths of the way from base 1 to base 2. This ratio is useful because it determines what voltage, as a percentage of the power supply voltage, is needed to turn on the UJT. This will be made clear shortly.

As shown in Fig. 9-5, the UJT is electrically equivalent to a voltage divider with a diode connected to a point between the two resistors. The diode comes about from the fact that the semiconductor junction conducts more easily in one direction than in the other.

202

Input signal (represented by Ve) is applied between the emitter (E) and base 1 (B1). Voltage Vbb biases the bar for proper operation. This bias voltage is always connected so that B2 is positive with respect to B1. The voltage at point K on the bar from the common lead (B1) will be somewhere between 0 and +10v, depending upon the relative values of Rb1 and Rb2.   As long as Ve is lower than the voltage at Point K, the diode formed by the P junction is reverse biased, so that only a slight leakage current can flow in the emitter circuit. The UJT is then said to be off. The emitter leakage current is very small in a UJT, which is an important characteristic of this device.

When control voltage Ve becomes greater than the voltage at Point K, the diode is forward-biased and conducts freely. Current from Ve then flows through Rb1 and the emitter. The resistance of Rb1 drops to a low value when emitter conduction starts, so that the emitter current is high. The UJT is then said to be turned on. The on and off current paths are shown in Fig. 9-6.

The reason why the resistance of Rb1 decreases can be briefly summarized as follows. The semiconductor material forming the bar is only lightly doped, so that it normally has but few charge carriers; another way of saying that the bar has high resistance. However, when the emitter junction becomes forward-biased and conducts, the emitter injects additional charge carriers into the bar, so that the bar resistance decreases. The decreased resistance allows a larger emitter current to flow, which means that more charge carriers are being injected into the bar, still further decreasing the resistance. As a result of this accumulative or regenerative action (called negative resistance), the emitter current very quickly reaches a high value after the emitter voltage reaches a value high enough to turn the UJT on. Although the resistance of the upper section of the bar also

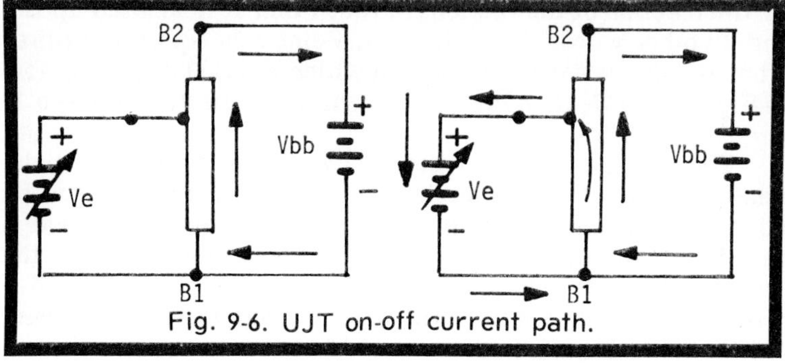

Fig. 9-6. UJT on-off current path.

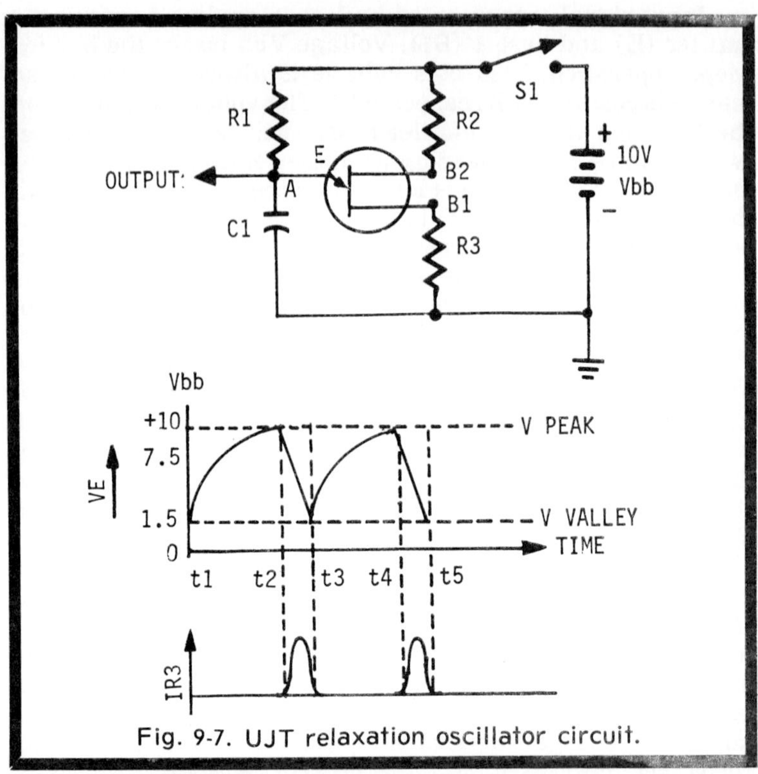

Fig. 9-7. UJT relaxation oscillator circuit.

decreases during the on period, by far the greater part of the reduction in resistance takes place in the lower section of the bar, since the emitter-injected charge carriers are mostly traveling within the lower section.

As mentioned before, the UJT switches on when control voltage Ve becomes greater than the potential at Point K. Since the resistance along the rod is evenly distributed when the UJT is off, the voltage at Point K is equal to Vbb multiplied by the fraction of the rod length that Point K is located up the rod. In other words, the voltage at Point K is equal to n times Vbb, and this is the Ve voltage at which the UJT turns on. If n in Fig. 9-5 is 0.75, then the device turns on when Ve reaches 0.75 times 10 or 7.5v.

The voltage at which the UJT switches on is called the peak voltage. Once the device is on, it will remain in the on condition as long as emitter voltage Ve remains above a minimum voltage known as the valley voltage. When the valley voltage is reached, the transistor switches off, and it will not turn on again until the peak voltage is again reached. The valley voltage is in the range of 1 to 2v for a typical circuit.

Thus we see that the unijunction transistor can be turned both on and off by means of the voltage from emitter to base 1. An important point to note is that the input resistance of the unijunction transistor is very high (of the order of megohms) while the transistor is in its off state. This is because the input is applied to a reverse-biased diode. However, when conduction begins, the input resistance becomes low, in the order of a few hundred ohms.

Pulse-Forming UJT Circuits

You can better understand the operation of the UJT by looking at how it operates as a relaxation oscillator, as shown in Fig. 9-7. When power is first applied to the circuit by closing switch S, the emitter is reverse-biased, so the transistor is cut off. Capacitor C1 then begins to charge through R1. Since the transistor is reverse-biased, it has a very high input resistance and does not load the capacitor.

The charge period is shown (Fig. 9-7) from time t1 to t2. At t2, the voltage from the emitter to base 2 reaches 7.5v and the transistor switches on. Capacitor C1 then discharges rapidly through the emitter-base 1 junction until the emitter base voltage falls to 1.5v (V valley), at which time the transistor switches off again. The capacitor then starts charging again, and the cycle repeats itself.

Resistor R3 is used to limit the emitter current during discharge to a value that the UJT can safely handle. Resistor R2 is used with sawtooth oscillator circuits for better temperature stabilization to reduce variations in the sawtooth frequency. When relatively high Vbb values are used, R2 is also needed to prevent thermal runaway.

If we take the output from Point A rather than Point B1, as in Fig. 9-8, we now have a pulse generator suitable for firing an

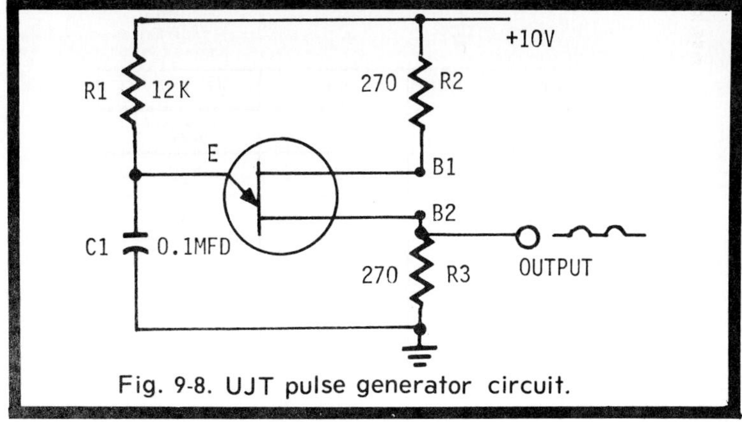

Fig. 9-8. UJT pulse generator circuit.

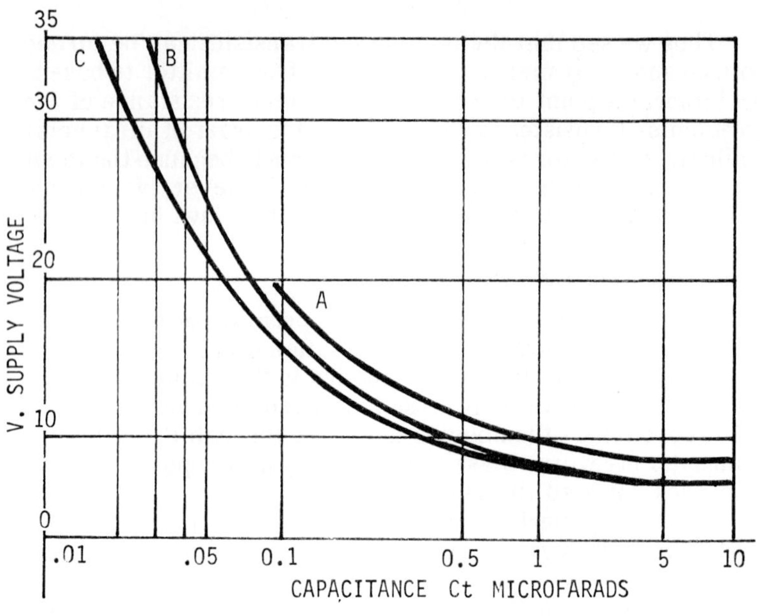

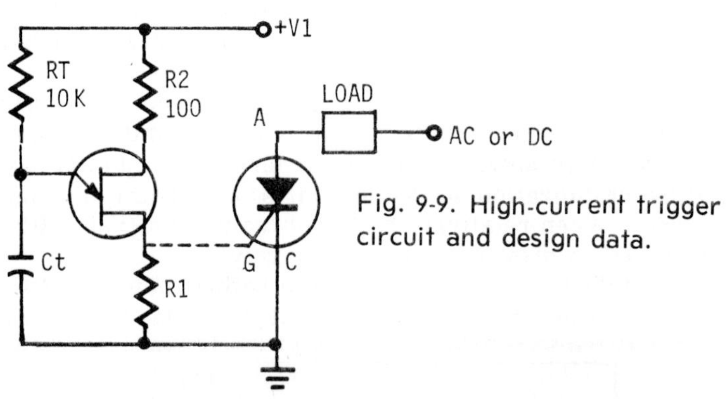

Fig. 9-9. High-current trigger circuit and design data.

SCR	CURVE	R1	V1 (MAX.)
C80	A	$27\Omega \pm 10\%$	20V
C60 (2N2073)	B	$27\Omega \pm 10\%$	35V
C55, C56	C	$47\Omega \pm 10\%$	20V
C45, C46 ZJ257 ZJ285	C	PULSE TRANS. PE2231	35V

SCR. While the transistor is cut off, very little current flows through R3. However, when the UJT switches on, the resistance of the bar becomes very low, and a relatively high current flows through the device. This current flows only during the brief period of time that capacitor C1 is discharging, so a pulse of current flows through resistor R3 (see Fig. 9-7). The surge of current flowing through R3 develops a high-amplitude voltage spike across R3. Resistor R2 is often omitted. Although this omission results in poorer frequency stability, it does give an output pulse of higher amplitude.

SCR TRIGGER CIRCUIT DESIGN

SCR trigger circuits using unijunction transistors are simple and compact with low power consumption, yet can easily trigger a 235 ampere SCR. An added advantage is that triggering is assured over a wide ambient temperature range. This assurance is available by using the trigger circuit outlined in Fig. 9-9. This circuit essentially merges the particular UJT and SCR in a configuration that takes into account the dynamic triggering requirement of the SCR.

Whereas most SCR spec sheets give consideration only to the static requirements, these design curves give the condition for the required minimum trigger energy. It is then simple to double the value of Ct or increase V1 for a guaranteed 2 to 1 above the minimum which is often desirable to decrease the SCR turn-on time and switching losses.

The value of R1 in the trigger circuits is kept low enough to prevent the DC voltage at the gate, due to interbase current, from exceeding the minimum gate triggering voltage for the SCR. The design of a suitable SCR trigger can be achieved easily and rapidly by using the design curves that are given. These curves gives the minimum supply voltage required to guarantee triggering of the various types of SCRs over the specific temperature range as a function of the UJT emitter capacitor, Ct. The value of resistor Rt is not important for the purposes of the design, provided that it is within the limits required for the UJT to oscillate.

If R2 is significantly greater than 100 ohms, the minimum supply voltage which is assumed (V1) should be calculated from the minimum supply voltage (V1) given in the design curves using the equation:

$$V1 = \frac{(2200 + R2)\ V1}{2300}$$

If two or more SCRs in parallel are to be triggered by a single UJT, the design of the trigger circuit must take into consideration the possibility that an SCR with a low gate resistance may be paralleled with an SCR having a high gate resistance, thus loading down the output pulse sufficiently to prevent the second SCR from being triggered. To reduce this possibility it is recommended that the trigger pulse be coupled by means of a separate capacitor to each SCR gate. Such capacitors act to equalize the charge coupled to each gate during the trigger pulse and thus tend to reduce the effects of unequal loading. The optimum capacitor value for this purpose has been found to be 0.1 mfd. In addition to this capacitor, a resistor having a value of 220 ohms to 1K should be connected between gate and cathode of each SCR. Using this approach (equalizing capacitors), the design curves in Fig. 9-9 can be used to parallel trigger an SCR, provided that the minimum supply voltage is multiplied by a factor of 1.5 if two SCRs are to be triggered in parallel and by a factor of 1.8 if three SCRs are to be triggered in parallel. Since the gates are not direct coupled, the maximum supply voltage allowed is limited only by the 35v rating of the UJT.

VARYING LOAD CURRENT BY PHASE CONTROL

Of course, one way to dim a light is simply to connect a rheostat in series with the lamp. If the lamp, or bank of lamps, uses several hundred watts or more, a big, clumsy rheostat is

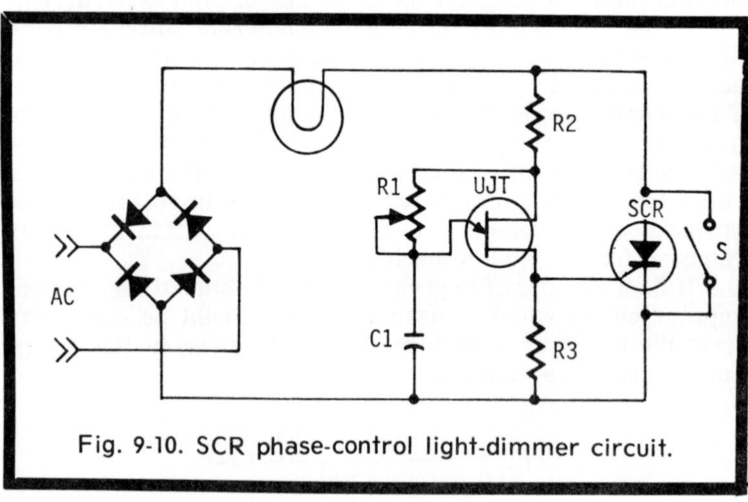

Fig. 9-10. SCR phase-control light-dimmer circuit.

involved in order to dissipate the heat generated. Also, you are paying for all the power wasted in the rheostat.

The use of an SCR as a light dimmer, as in the circuit shown in Fig. 9-10, is much more efficient and generally more practical. Dimming is accomplished by limiting current flow through the lamp to only part of each cycle, so that the average current through the lamp is reduced. This method of regulating power to a load is called phase control.

Because of bridge rectifier action, full-wave rectified current as shown in Fig. 9-11 flows through the lamp throughout the cycle if conduction is allowed for the entire cycle. This will be the case if switch S is closed, cutting out the SCR circuitry. The lamp then burns at full brilliancy. The AC supply voltage is rectified so that the single SCR can conduct on both half-cycles. The fact that pulsating DC rather than AC is flowing through the lamp has no effect on the brilliancy.

If switch S is open and the SCR fires at the beginning of each cycle, the continuous current flow shown in Fig. 9-11A will be obtained. If we delay the firing of the SCR until the middle of each cycle, as shown in Fig. 9-11B, current flows through the lamp only during one half of each cycle. Remember that the SCR cuts off when the anode current drops to near zero at the end of each cycle, and it does not start again until the SCR is fired. The average power delivered to the lamp will be half as much as before, and the lamp will not burn nearly as brightly. By triggering the SCR still later in the cycle, the lamp can be dimmed still more. By varying the firing time, any degree of brightness from full brilliancy down to complete darkness can be obtained.

Now the problem that arises is how to trigger the SCR at exactly the 90 degree point of each half-cycle, or at any other time required by the brilliancy desired. For satisfactory control of the firing time a pulse, such as that obtained from a unijunction transistor circuit, must be used. Since the positive voltage at which the SCR fires varies so much with temperature, we must have a sudden, sufficiently sharp rise in

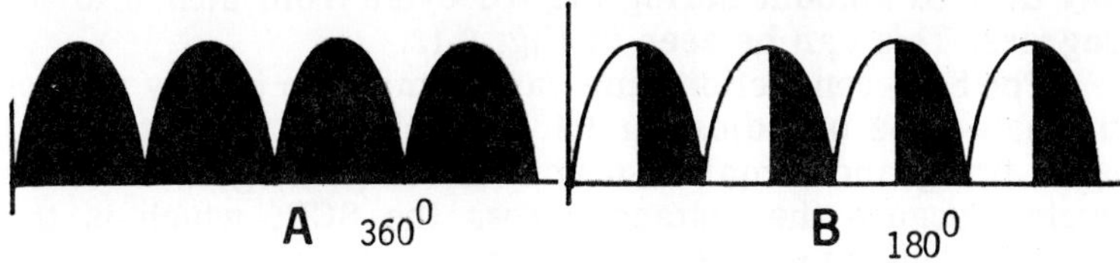

Fig. 9-11. Phase-control circuit current flow.

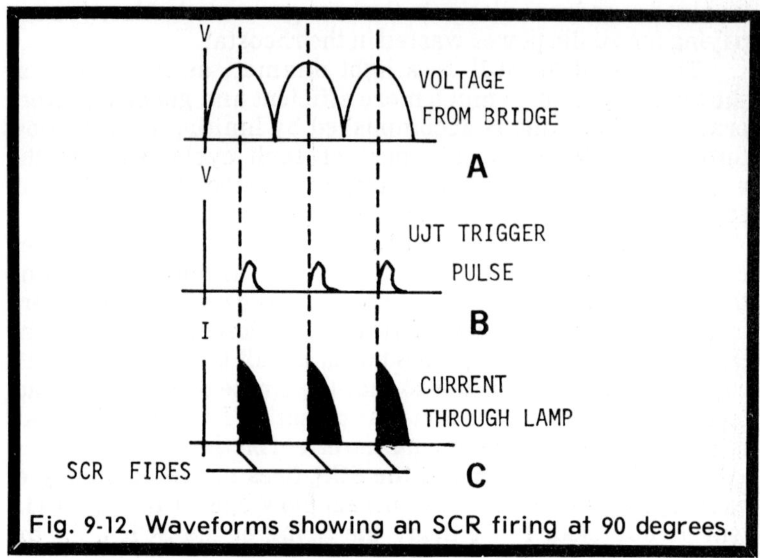

Fig. 9-12. Waveforms showing an SCR firing at 90 degrees.

gate voltage (such as the sharp pulse output from a unijunction transistor) to assure firing at any temperature.

By varying the setting of potentiometer R1, Fig. 9-10, the UJT is adjusted to emit a triggering pulse at the desired time within each AC cycle. For firing at 90 degrees after the beginning of each cycle, the waveforms and current flow are as shown in Fig. 9-12.

The UJT pulse circuit used to develop the trigger in Fig. 9-12 is the same as that shown in Fig. 9-8, except that instead of being connected to the DC Vbb supply, timing resistor R1 and base 2 are connected through a current-limiting resistor (R2) to the pulsating voltage line. The operation of the circuit is the same as that of the DC version; i.e., the capacitor is charged through R1 and causes the UJT to turn on and generate a pulse when the voltage at the emitter of the transistor reaches the peak voltage. By adjusting R1, the UJT turn-on can be delayed any desired amount during the AC cycle from near 0 to 180 degrees. This can be seen in Fig. 9-13.

The SCR conduction time can be made to be any part of the half-cycle by adjusting R1. Capacitor C1 is discharged upon firing and remains so until the beginning of the next cycle, because the voltage across the SCR, which is the voltage applied to the R1-C1 timing circuit, drops to near zero and stays there as long as the SCR is conducting. At the end of the cycle, SCR conduction stops and the voltage applied to R1-C1 is the voltage output from the bridge rectifier. While we use

210

the basic circuit in Fig. 9-10 for dimming a light, it could be used for any application in which it is necessary to vary the load current. Rather than using a bridge rectifier (as in Fig. 9-10) to achieve full-wave operation, you could connect another SCR in parallel with the first, but with opposite polarity. Each SCR would then be in use on alternate half-cycles.

SCR Motor Control

One of the most important uses for the SCR circuit in Fig. 9-10 is as a speed control for a series DC or a universal AC motor. Just as the brightness of a lamp can be increased or decreased by controlling the power to it, so the speed of a series motor can be increased or decreased by controlling the current to it.

Power economy is not the only advantage of using an SCR for motor control; other advantages are more important. With the phase control method of varying speed, the controlled motor has higher torque at low speeds than it does if a rheostat is used in series with it. Higher torque means higher armature twisting power. This is often of the greatest importance.

As an example, let's compare two sewing machines, one using a rheostat for speed control and the other using an SCR. If we want to sew as slowly as possible, we start the rheostat-controlled machine with maximum series resistance and cut down the resistance until the motor starts. A sewing machine requires quite a lot of torque (armature twist) to get it started turning, but it runs much easier after it is once started. With rheostat control the armature has little torque, so that the

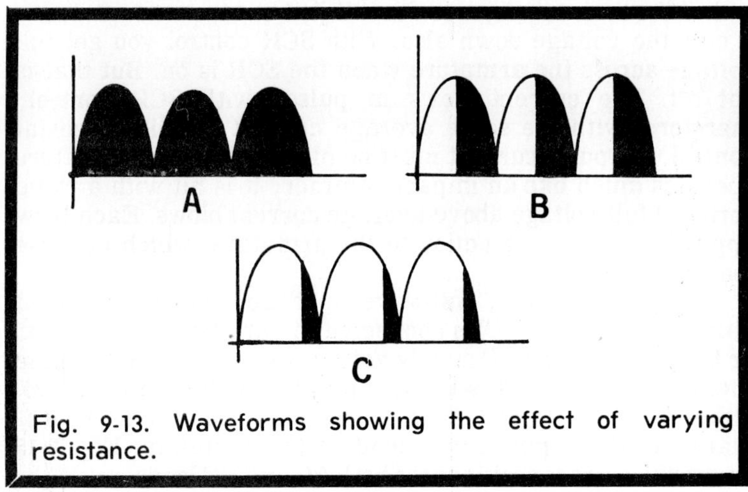

Fig. 9-13. Waveforms showing the effect of varying resistance.

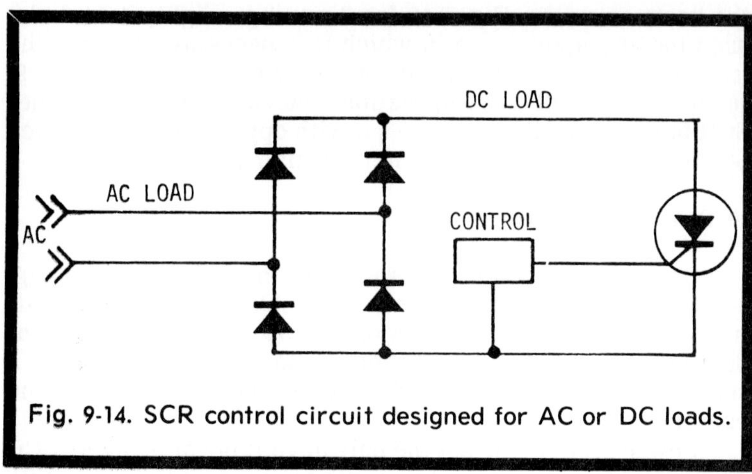

Fig. 9-14. SCR control circuit designed for AC or DC loads.

series resistance must be cut down considerably to get enough current through the armature to start the machine. Then, after it is started, the machine turns easier and takes off at a much higher speed than desired because of the relatively high armature current. Now you must increase the series resistance to slow the motor down to the desired speed, but by the time you have done that, you may have ruined the material being sewed. With the SCR-controlled machine, only a little current through the armature develops a lot of twisting power, and thus gets the machine going. It will continue to run slowly after starting because the armature current is low.

The reason you have such low torque with rheostat control is that the series resistance doesn't just cut the current down, it cuts the voltage down also. With SCR control you get full voltage across the armature when the SCR is on. But that is not all. The current flows in pulses with SCR control; therefore, with the same average current as with rheostat control, the pulse current must be higher. Thus the armature operates much like an impact hammer; it is hit with a rapid series of full-voltage above-average current blows. Each blow applies high twisting action to the armature, which gets the load moving.

As shown before, full-wave operation can be obtained either by using two SCRs connected in parallel, back-to-back, or by rectifying the AC supply voltage as in Fig. 9-14. With the method shown in Fig. 9-15, the circuit can be used to control either a DC or AC load. If an AC load is to be controlled, it is placed in the supply line ahead of the rectifiers. Fig. 9-15 shows the correct positions for both AC and DC loads.

212

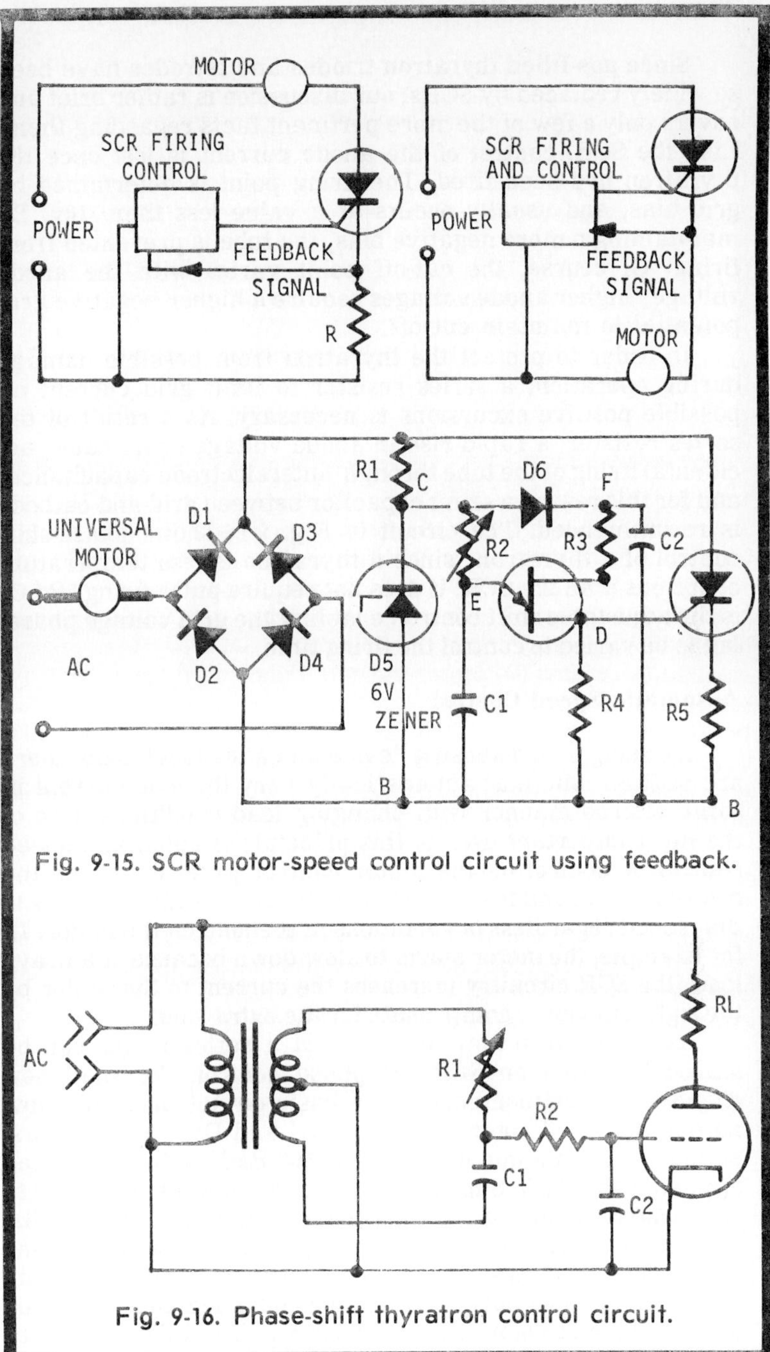

Fig. 9-15. SCR motor-speed control circuit using feedback.

Fig. 9-16. Phase-shift thyratron control circuit.

Thyratrons

Since gas-filled thyratron triodes and tetrodes have been so widely replaced by SCRs, our discussion is rather brief and covers only a few of the more pertinent facts regarding them. Like the SCR, control of the anode current is lost once the thyratron has been fired. The firing point is determined by grid bias, and usually occurs at a value less than -10v. By maintaining a more negative bias, the tube is prevented from firing. Of course, the cut-off point varies with the anode voltage; higher anode voltages require a higher negative grid potential to maintain cut off.

In order to protect the thyratron from possible damage during operation, a series resistor to limit grid current on possible positive excursions is necessary. As a result of the series resistor, a rapid rise in anode voltage could cause accidental firing of the tube through interelectrode capacitance, and for this reason a shunt capacitor between grid and cathode is recommended. The circuit in Fig. 9-16 shows phase-shift control of a thyratron; since a thyratron is less temperature conscious than the SCR, it does not require pulse firing. R1-C1 is the adjustable shift control, enabling the grid voltage phase-lag to be varied to control the firing time.

Automatic Speed Control

By using a load-sensing device and a feedback loop, there are SCR circuits that automatically vary the load current in some desired manner with changing load conditions. One of the most important uses of this principle is automatic speed control. A control on the motor control panel is set for the desired speed, and the control circuitry then holds the motor to that speed regardless of variations in the load on the motor. If, for example, the motor starts to slow down because of a heavy load, the SCR circuitry increases the current to the motor by the right amount to compensate for the extra load.

You have seen how the speed of a series motor can be varied by using an SCR for phase control. We shall now discuss some refinements on the basic circuit that will allow automatic speed control of universal and DC series motors. Suppose we had a motor running at 500 RPM with a light load applied to the shaft. Suppose also that in order for the motor to run at 500 RPM the current pulses through the SCR looked like the pulses in Fig. 9-13A. Now, let's apply a heavier load to the motor. The motor speed will then decrease. In order to bring the speed up to its normal value, we could decrease the value of the timing resistor in the UJT circuit. The UJT would then fire earlier in the cycle, and the current through the motor might then look like Fig. 9-13B.

Now let's reduce the load to a value less than it originally was. The motor speed will increase to higher than 500 RPM. To compensate we could decrease the motor speed to 500 RPM again by increasing the value of the timing resistor. The current through the motor might then look like Fig. 9-13C.

So we see that two things are necessary in order to keep the motor speed constant under varying loads: (1) some method of detecting any change in speed and (2) circuitry for feeding back a signal (called an error signal) from the speed-change detector to the SCR control circuitry in such a manner as to vary the conduction time of the SCR.

Two methods of detecting motor speed change are in common use. The feedback error correction signal in Fig. 9-15 is the voltage drop across the resistor, which is in series with the motor. If the motor speed changes, the value of the motor current changes; thus the feedback voltage signal changes. The feedback signal voltage acts through the SCR firing and control circuitry to vary the firing time of the SCR, thus slowing down or speeding up the motor as required to bring the speed back to the proper value.

Speed Control by Sensing Motor Current Change

Fig. 9-15 is a complete circuit for an SCR motor-speed control that uses a method of load current sensing to control the SCR conduction time. The circuit is very similar to the full-wave bridge circuit discussed earlier, with just a few differences. First of all, there is a resistor (R5) in the SCR cathode lead through which the motor current must flow. It is the current flow through this resistor that provides the feedback. In addition, there is a 6v zener diode connected to Point C that prevents the voltage at that point from rising above +6v.

Clamping the Vbb supply to 6v has two advantages. First, it provides a smoother adjustment of the UJT timing circuit because the base 2 voltage is held constant at +6v while C1 is charging. Without a constant Vbb supply, both the B2 voltage and the voltage across C1 rise during the first quartercycle, but during the second quartercycle the B2 voltage drops while the C1 voltage is still rising. A more even control with R2 can be obtained if Vbb is held constant while C1 voltage is still rising. The second reason for clamping the voltage at Point C is to keep the base 2 voltage at a low value. Remember that the UJT firing voltage is low when the base 2 voltage is low. Since

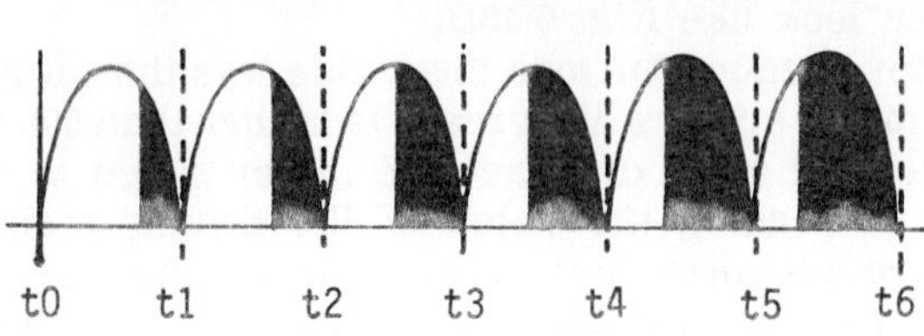

Fig. 9-17. Waveform depicting SCR current pulses.

the feedback voltage developed by the current flow through R5 is low, the UJT firing voltage must be kept low in order to work effectively. Diode D6 and capacitor C2 are added to the UJT circuit to keep the Vbb up to +6v during the conduction time. When the voltage between A and B is greater than 6v, D6 conducts and charges C2. The voltage from B to C cannot rise above 6v because of the zener diode. But when the voltage from B and A decreases below 6v, which it will when the SCR conducts, diode D6 will be reverse-biased and C2 will hold Point F at 6v. In this manner, C1 is allowed to charge to the voltage between Points A and B during SCR conduction without having the UJT fire again.

When the line voltage is cut in, the basic operation parallels the preceding explanation with the Vbb supply clamped at +6v by the zener diode. Since the unijunction has a value of 0.67 and as the voltage at Point E reaches +4v, the UJT must turn on and provide a triggering pulse to cause the SCR to fire. Since the SCR then conducts for the remainder of its cycle, as shown in Fig. 9-17, the voltage from Point B to A during SCR conduction falls to a value equivalent to the drop across the series connection of the SCR and R5. Should the current through the motor be small at the time, one ampere would offer very little voltage drop across R5's low value; the B to A voltage would be very little more than the 1v drop of the SCR.

With the motor running at its normal speed, by increasing the load the motor slows down in proportion to the load increase and also must draw more current as a result of the reduction in the counter EMF developed. Operation now continues in exactly the same way as before, only during conduction the Point B to A voltage increases as more current is flowing through series resistor R5. With the UJT turned off, having generated its firing pulse, timing capacitor C1 charges to voltage across A and B during conduction. However, the UJT will not fire because the base 2 voltage holds at +6v while capacitor C2 is charging. C1 discharges during conduction to a

216

voltage proportional to the motor current, and as a result of this residual charge the time required to reach a +4v charge on the next cycle is much less. During the next half-cycle, the UJT offers its pulse earlier and thus permits current flow through the SCR for a longer portion of the cycle. This causes more power to be delivered to the motor, increasing its speed again as it endeavors to maintain a constant rate. As the load is increased more, the voltage across A and B increases, and since even more current is needed by the motor, a higher residual charge is left on the capacitor to allow it to charge to +4v even sooner. This permits the UJT to pulse sooner and trigger the SCR earlier for a longer delivery of power in an effort to establish the constant rate of motor speed desired under greater load requirements.

The two necessary steps for automatic speed regulation are provided by the circuit of Fig. 9-15. It (a) senses a change in speed as a result of a change in current, and (b) the SCR conduction time is varied by the triggering pulses applied from the UJT, which in turn is controlled by the residual charge on the timing capacitor.

Regarding the value of series resistor R5, which is quite critical, insufficient correction in speed results if the resistance is too small, and overcorrection if too large, causing the motor to oscillate or hunt. The adjustable type is preferable to permit the best regulation under various conditions.

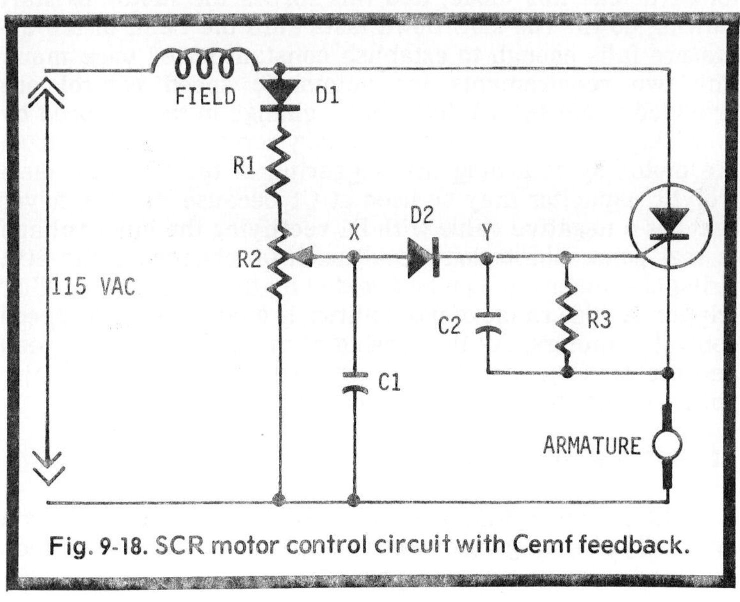

Fig. 9-18. SCR motor control circuit with Cemf feedback.

Although a series AC motor may be controlled satisfactorily by feedback based on the current flow through a series resistor, the counter EMF developed by the motor armature presents another method of feedback control worthy of consideration. This form of feedback, resulting from the generator action of the motor, varies according to the speed of the armature. The faster the armature turns, the greater the counter EMF as a direct result. By measuring the Cemf presented by the armature, a positive indication of motor speed is available. The diagram in Fig. 9-18 shows a method of connecting an SCR and a pair of diodes to control motor speed utilizing the counter EMF produced by the armature.

The SCR is cut off until the voltage Point X builds up enough to forward bias D1, causing D2 to conduct which switches on the SCR. The wiper on R2 adjusts the motor speed by regulating the time necessary for the voltage at X to rise high enough to forward bias D2. The motor runs at a predetermined speed, and as the load is reduced, the motor speed increases as too much power is being delivered to the smaller load. The increased motor speed results in more Cemf voltage generated across the armature.

To forward biasing D2 the voltage at Point X must overcome the Cemf fed to Point Y through C2. For this reason, it now takes longer for the voltage at X to increase enough to forward bias this diode, and this forces the motor to start slowing down. The slow down lasts until the Cemf of the armature falls enough to establish constant speed once more. The two requirements for automatic speed control are provided again by (a) detecting a change in motor speed by the rising Cemf, and (b) the necessary correction to slow down the motor by retarding the triggering of the SCR. An electrolytic capacitor may be used at C1 because Point X never reaches a negative value with D1 rectifying the input voltage to that point. Limitations resulting from changes in the SCR firing characteristic can be avoided by the addition of a UJT trigger. A wide range of possibilities is available for the speed control of motors, but the sensing of any fluctuation in speed and the development of corrective measures are the major points in the design of such circuitry.

SENSITIVE SWITCHES

The circuit in Fig. 9-19 switches a load in response to a gradually changing signal such as light on a cadmium sulfide

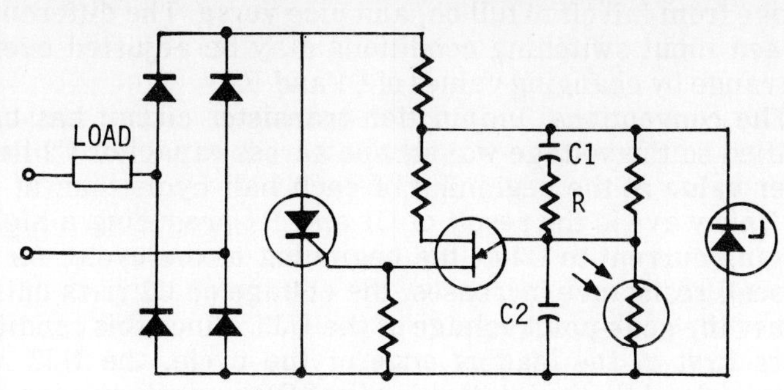

Fig. 9-19. Sensitive AC power switch circuit.

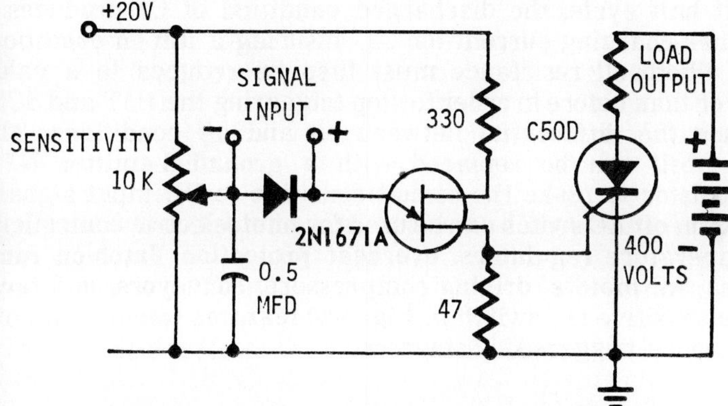

Fig. 9-20. Sensitive DC power switch circuit.

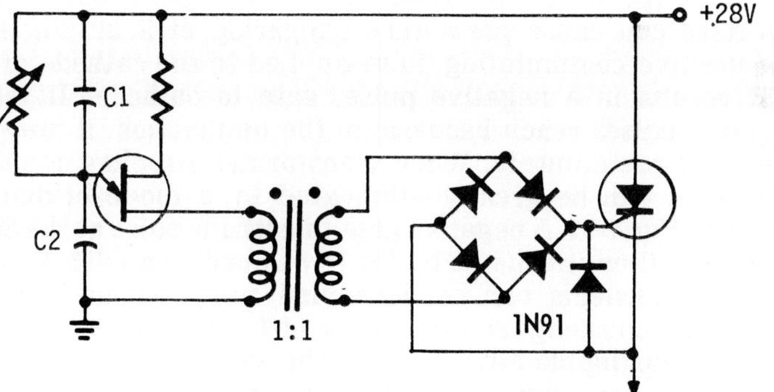

Fig. 9-21. Trigger circuit designed to eliminate the effects of transients.

photocell, temperature of a thermistor, etc. Switching is positive from full off to full on, and vice versa. The differential between input switching conditions may be adjusted over a wide range by changing values of C1 and R1.

The conventional unijunction transistor circuit has been modified so the voltage waveshape across capacitor C2 has a higher value at the beginning of each half cycle than at the end. This wave is the result of C1 and R1 producing a higher charging current to C2 at the beginning of the cycle. As the photocell resistance increases, the voltage on C2 rises until it reaches the peak-point voltage of the UJT. Since this condition occurs first at the leading edge of the cycle, the UJT will trigger only at that point, turning the SCR on early in each half cycle. Triggering the SCR removes voltage from the UJT circuit, and capacitor C1 discharges. At the beginning of the next half cycle, the discharged condition of C1 produces a higher charging current for C2, assuring a full-on condition.

Photocell resistance must then be reduced to a value lower than before in order to stop triggering the UJT and SCR, hence the differential between on and off conditions. The photocell can be replaced with a grounded-emitter NPN transistor to make the circuit sensitive to DC input signals. This on-off AC switch can be used for photoelectric controllers, temperature regulators, overheat protection, latch-on functions, AC motors, driving compressors, conveyors, and fans. The sensitive DC switch in Fig. 9-20 features latching action; i.e., once triggered, it stays on.

TRIGGER CIRCUIT TRANSIENT PROBLEMS

The impulse commutation often used in DC choppers and inverters can cause premature triggering of a unijunction. The positive commutating pulse applied to the cathode of an SCR results in a negative pulse, gate to cathode. If these negative pulses reach base-one of the unijunction, it may be triggered prematurely. When transformer coupling is used, transients can be greatly attenuated by a diode bridge as shown in Fig. 9-21. A negative gate-to-cathode pulse at the SCR may be further attenuated by the low-impedance 1N91. Supply voltage transients can be decoupled from the unijunction emitter by choosing the impedance of C1 and C2 so the transient has negligible effect on Vp. The sum of C1 and C2 constitutes the total unijunction timing capacitance.

The triggering circuit in Fig. 9-22 uses the UJT to drive the 2N526 from cut-off to saturation. Since we are not using the energy in C1 to trigger the SCR, a smaller .01-mfd capacitor

can be used and thus achieve UJT operation to 20 kHz. The
1N4154 keeps the emitter-base junction of the 2N526 reverse
biased except for the discharge interval of C1. R5 limits the
voltage amplitude of the trigger pulse to the SCR. The 2N526
also isolates the SCR turn-off pulse from the UJT timing
circuit. The square-wave inverter drive circuit has been used
successfully as the control and trigger source for parallel
inverters using C40 SCRs.

Pulse Suppression or Suppressors

Excessive sparking is frequently a problem facing the
designer of electronic control and sensing equipment as used
in manufacturing. Due to the Cemf generated by inductive
type circuitry (i.e., relays, chokes, and transformers),
sparking between contacts and actual puncturing of insulation
is encountered. This may readily be eliminated by a capacitor
and resistor in series across the contacts or switches so involved.

SWITCHING DEVICE SUMMARY

Diac: a double PNPN switch, or two-lead AC switch
having high resistance to voltage applied in either direction
until a critical voltage is reached (35v). As this breakdown
voltage is reached, the diode switches to a lower resistance
value and causes the voltage to drop to a much lower value. It

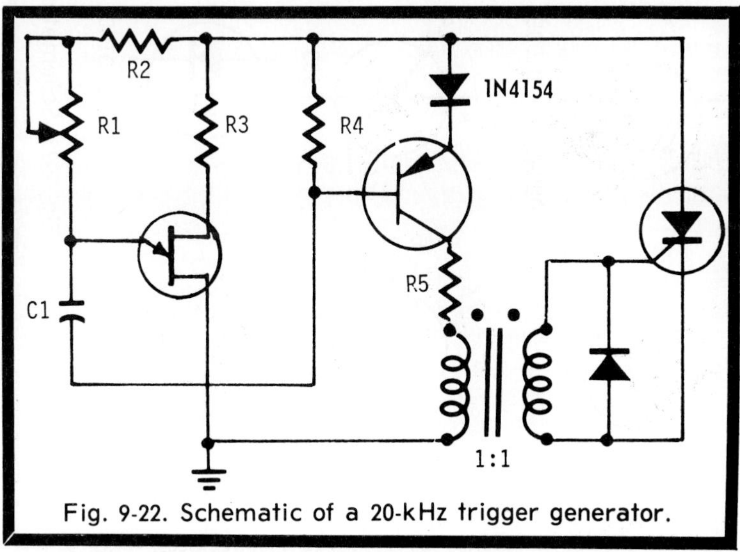

Fig. 9-22. Schematic of a 20-kHz trigger generator.

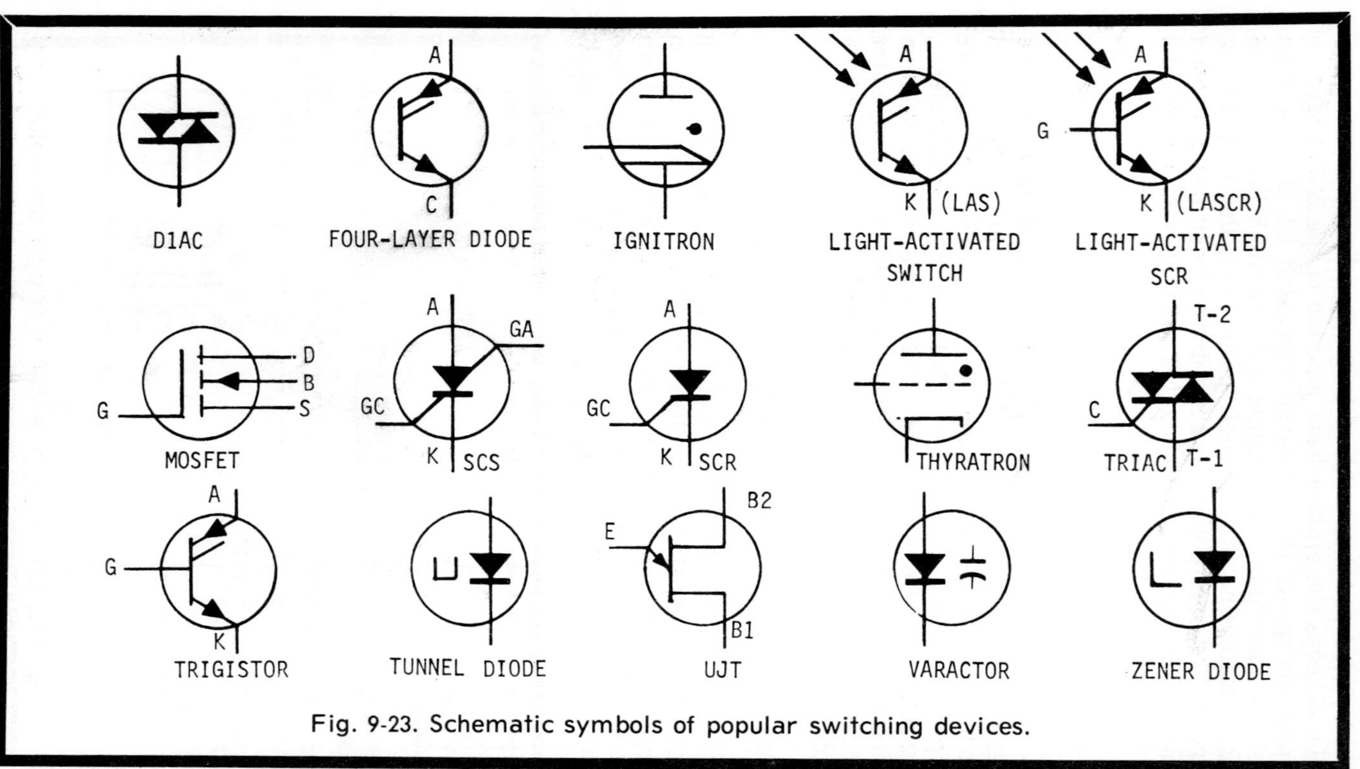

Fig. 9-23. Schematic symbols of popular switching devices.

differs from the 4-layer and zener diodes by operating equally well in either direction, and its most frequent use is as a triggering diode for a Triac.

Four-layer diode: Shockley or PNPN diode, a 2-terminal switching device permitting a very small current to flow in a forward direction with forward-bias until its critical voltage is reached. The diode switches into high conduction when this firing voltage is reached, and continues to conduct freely until current is reduced to below a minimum holding value. Here it switches back to the off state; it differs from a zener because the voltage across it drops to a very low value as soon as conduction begins.

Ignitron: Mercury vapor rectifier capable of handling high-power. It has a small pool of mercury inside the glass envelope that is ionized by applying a voltage to the ignitor. As with other rectifiers, the ignitron conducts only when its anode is positive as opposed to its cathode. It is capable of handling several thousand amperes as required in welders, and is often controlled by a thyratron.

Light-activated diode switch: (LAS) Similar to the light-activated SCR without the gate terminal, and may be turned on only by a light beam.

Light-activated SCR: (LASCR) A low-power unit similar to the SCR, and may be triggered by a light beam in its glass window as well as by a regular gate signal. Once triggered, anode current is independent of light or gate triggering until the anode current is interrupted momentarily, either by switching off or reversal.

Silicon controlled switch: (SCS) A PNPN transistor or NPNP switch. A low-power switching device, somewhat similar to the SCR but more sensitive; it can be turned off by a negative signal applied to the gate, but it must be of greater amplitude than positive signal needed to turn it on. It has a fourth lead that can be used to eliminate accidental turn on.

Silicon controlled rectifier: (SCR) controlled rectifier, thyristor, Trinistor, GCS. A high- or low-power rectifier according to type desired. It switches on with the anode positive with respect to the cathode, and only by applying a trigger to the gate terminal. Turn off is accomplished only by reducing the current through it to less than the minimum holding value.

Thyratron: A 3-element gas-filled tube with power-switching capability and similar to a mercury vapor rectifier. It utilizes a control grid to turn on or hold off. Though lacking in many ways, it is similar to the SCR basically. Thyratons are being replaced in equipment of late design by the SCR, although they are still in use in older control systems.

Triac: A symmetrical switch or tri-lead AC switch. A power switching device that acts like two SCRs in parallel; it will conduct in either direction when triggered. Triggering may be alternately positive and negative with the load current; negative offers more reliable triggering. The common terminal (T1, Fig. 9-23) corresponds to a transistor's emitter, or a tube's cathode. Actually, the Triac resembles a Diac with a gate.

Trigistor: Transwitch or binistor. A low-power unit bearing a similarity to the SCS except that a lower power at the gate will turn it off. It is used as a flip-flop and is represented by the same symbol as the SCR. The binistor really belongs to the transistor group, rather than the SCR, and it is closer to the SCS in operation. It has an additional control lead, called the injector, with operating voltages somewhat lower than those of SCS devices.

Tunnel diode: Esaki diode. Operates in a different manner than the usual diode. The voltage changing from peak to valley causes the current through the diode to switch from high to low. Thus the tunnel diode has exceptional switching capability, and operates at very high switching speeds. It is also used as an oscillator, but since it has only two terminals, it is much more difficult to control than a transistor.

As the forward bias voltage rises from zero, current increases rapidly until the peak forward current is reached (as well as peak voltage). Increasing the voltage now causes the current through the diode to decrease, a condition known as negative resistance. Current continues to drop as the voltage increases until the valley current point is reached; the voltage at this point is called valley voltage. Passing this point, the current again increases with voltage.

224

Chapter 10

Probe Problems

Before we consider a variety of switching devices not yet covered, let's look at several applications found in air and marine systems.

AVIATION ELECTRONICS

Electronic systems provide aircraft with navigation, guidance, communication, and numerous warning and control systems. The automatic control system or auto-pilot makes flying supersonic aircraft possible, as well as relieving the weary pilot on long trips and in bad weather. In this system, a complex servo-system relieves the pilot by constantly correcting the pitch, roll, and yaw attitudes of the aircraft. Gyro scopes provide signals to control corrections made for pitch and roll. The information signals are fed into an amplifier, either a solid-state or magnetic type, with an output feeding an actuator. The actuator moves the control surfaces until the proper flight path is restored. By using feedback-loop techniques, overshoot is prevented when large signal information would tend to produce quick corrections, instead of the desired slow, steady control. The instrument landing system (ILS) permits landing in poor visibility with reasonable safety.

Navigation Control

The ADF, or automatic direction finder, is widely used for navigation. By tuning to a fixed-location radio transmitter, the pilot sets his radio compass. A comparison of the resulting bearing with another station or the magnetic compass can easily be used to determine an exact position. The ADF

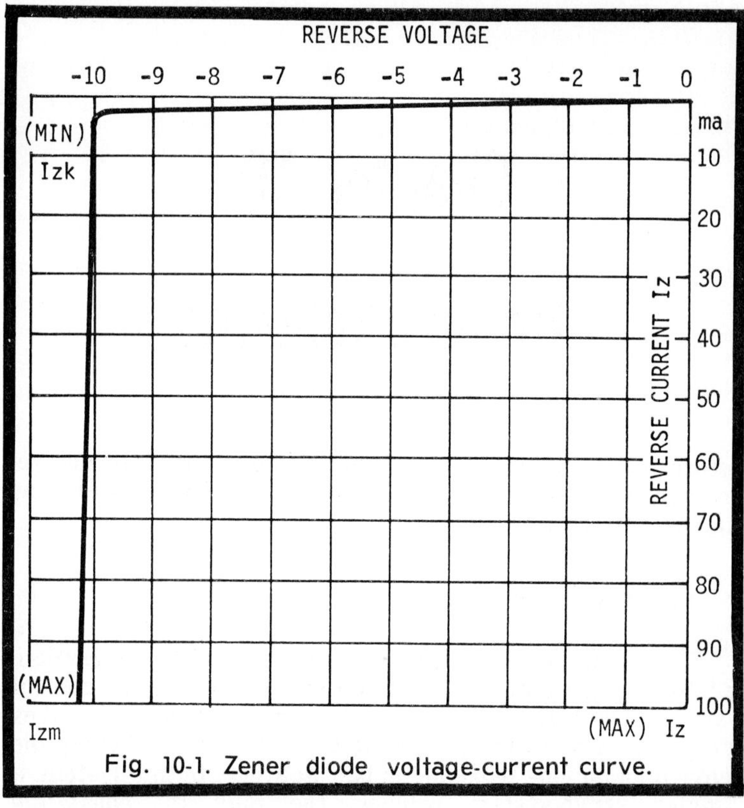

Fig. 10-1. Zener diode voltage-current curve.

system employs the characteristics of the common loop antenna, and by adding a small vertical antenna to reduce ambiguity, the system is quite reliable.

Sense Antenna

A phase-sensitive antenna is used in conjunction with a loop in ADF systems to provide positive information as to the direction of travel to or from a ground station. Installed near the loop, the sense antenna voltage is maximum at the same time the radio signal is maximum and zero when that signal is zero. Its induced voltage is in phase with the transmitted signal, while the loop shows the greatest voltage difference as the radio wave is passing through zero and no difference occurs as the wave reaches its maximum level. Thus the loop antenna is exactly 90 degrees out of phase (from a voltage standpoint) with the sense antenna.

If the audio wave comes from the left, it induces voltage in

226

side A and then C, and lags the sense antenna by 90 degrees. When the wave arrives from the right side, the polarity is opposite and now leads the sense antenna by 90 degrees. By shifting the loop voltage an additional 90 degrees and combining it with that from the sense antenna, it will either be in phase or 180 degrees out, depending on the direction. By adding the two patterns algebraically, a cardioid pattern is obtained. Since a cardioid pattern has only one maximum and one minimum response point, it clearly points out the direction of the transmitter from the aircraft. By selecting the point of minimum response, greatest accuracy is assured, and this is done automatically from the ADF receiver through a servo.

The loop antenna bearing information is shown on the panel indicator in the cockpit. A digital counter displays the selected station's frequency in kHz and the bearing indicator, using two pointers, provides simultaneous bearings from two receiver systems.

Loran

Although many navigational aids are very useful for distances reasonably close to the transmitted signal, long-range navigation (LORAN) is a system that pinpoints a position on transoceanic and other long-range flights. Developed during WW2 for military needs and currently used for commercial flights over the Atlantic and Pacific oceans, loran establishes an aircraft's position by measuring time difference of radio pulses arriving from two or more pairs of transmitters. Since each pair of stations sets up a position, the point of intersection indicates the exact location of the aircraft.

The main transmitter sends out a series of uniformly spaced pulses in all directions, and the slave transmitter, a known distance away, sends similar pulses triggered by the main station. Sometimes a single master station may trigger two slave stations and this is known as double-pulsed transmission. By synchronizing his loran with the PRR of the master and slave A, and again with the master and slave B, the pilot's position is the point where the two readings intersect on the map. The PRR is the identifying mark of the loran station; the number of pulses transmitted per second establishes identity. The slave station receives a pulse from the master station and rebroadcasts it after a predetermined delay.

As you can see, pulse and switching applications touch virtually every aspect of human activity, and the horizons are ever widening. Before taking up additional uses, let's consider in more detail some of the components mentioned just briefly in Chapter 9.

TUNNEL DIODE

A tunnel diode differs from other diodes in that it may be used to amplify or to oscillate. As such it has many advantages even over the transistor. It may be used at much higher frequencies than the usual transistor and it is not sensitive to temperature changes. It's lighter in weight, smaller in size, requires less power, and last but far from least, it offers a much lower noise figure, making it suitable for diverse applications. However, there are disadvantages which will prevent the tunnel diode from taking over the field at least for the present. Since it has only two terminals the tunnel diode is restricted in use for the moment and circuit design is a bit of a problem.

Since the charge carriers tunnel through the barrier potential rather than over it, as is the case with the usual diode, a tunnel diode requires much less energy. The tunnel barrier is exceptionally thin and charge carriers approaching the barrier quickly disappear as another reappears on the other side. The capacitive effect, so disruptive in the operation of ordinary diodes at high frequencies, is entirely lacking in this type. Therefore, tunnel diodes are capable of operation to several thousand MHz.

The negative resistance factor makes it possible for the tunnel diode to function as an amplifier or oscillator. While ordinary diodes require an increase in bias voltage for an increase in diode current, in the tunnel an increase in bias voltage over a certain range results in decreasing diode current. By the same token, the current may be increased by lowering the bias voltage. This characteristic is called negative resistance and permits any device so characterized to amplify and also to function as an oscillator.

ZENER DIODES

Zener diodes, avalanche diodes, or silicon regulators are commonly used as regulators or voltage limiters. They make use of the reverse-voltage breakdown and avalanche-current principles. Increasing the reverse voltage across the diode increases the number of minority carriers, because the extra voltage provides additional energy for breaking covalent bonds. Hence, the reverse resistance of the diode decreases with the increased voltage. Also, the high voltages increase the speed at which the charge carriers move. When the voltage becomes high enough, the charge carriers move fast enough to knock apart other covalent bonds, making additional charge carriers. The newly formed charge carriers help break more bonds, so that the effect is cumulative. The result, a very rapid increase in reverse current after a certain reverse voltage, called the breakdown voltage, is reached. This is called the avalanche current.

Zener diodes differ from ordinary diodes in that they are specifically designed for sudden breakdown when the designed reverse voltage is reached. Forward biasing need not be considered because zener diodes are not normally used with forward biasing. The curve in Fig. 10-1 shows that the reverse current is close to zero until the reverse bias voltage reached 10 volts. When this voltage is reached, the zener breaks down and becomes a conductor for any voltage above breakdown. Once the breakdown is reached, any increase in voltage does not increase the voltage across the zener but merely causes the current through the zener to increase rapidly. For this reason, we always use a series resistor with all zener diodes to prevent their destruction.

The important specifications are: Ez, zener breakdown voltage; Izk, minimum zener current for good regulation; and Izm, the maximum safe current that the zener can handle. Sometimes Izm may not be listed, but it can be computed from the wattage rating of the zener diode, using the formula:

$$Izm = P/E$$

The maximum safe current for a 50-watt zener diode with a breakdown voltage of 11 volts is:

$$50W/11V = 5 \text{ amperes}$$

Minimum Value of Izk

An important characteristic of the zener diode is the minimum value of zener current, Izk, the value of zener current at the knee of the curve in Fig. 10-1. This curve shows that there is a minimum current that must flow through the zener for the voltage across the zener to be a nearly constant value, Ez. If the current through the diode is allowed to drop below Izk, the voltage across the zener will be less than Ez. The value of Izk is usually a small fraction of the maximum safe zener current, Izm, and can be found on the curves available from the manufacturer for any specific diode. The value of Izk is generally on the order of a few milliamperes for zeners with an Izm value of about 1 ampere.

Zener Diode Voltage Regulator

You have witnessed how the voltage across a zener diode is held constant with variations in current flow through it. This property is very useful as a voltage regulator, and in Fig. 10-2, a 6-volt zener diode is used to obtain a constant 6 volts across the load (RL) for variations in input voltage between 8 and 10 volts.

In addition to maintaining a constant output voltage for variations in input voltage, the zener also holds the voltage constant for changes in load current; that is, for changes in load resistance RL. For varying values of RL, the zener diode resistance automatically adjusts itself to keep the voltage across it constant. But, it can do this only as long as the current through it remains above the value of Izk.

We can better understand by an example how the voltage remains constant across a zener diode while the load current varies. Let's say, for example, that input voltage E1 is constant at 8 volts, that Izk is 5 ma for the zener we are using, and that resistor Rs was so chosen that the current through the diode is 35 ma with no load resistor connected. As long as E1 remains at 8 volts and E2 is 6 volts, the current through Rs will be 35 ma. If we connect a resistor (RL) in parallel with the zener, part of the 35 ma current will flow through the zener and part through RL. The voltage across RL will remain at 6 volts as long as the portion of current flowing through the diode remains above Izk, which is 5 ma. In other words, we could draw up to 30 ma through RL and the voltage across it would remain 6 volts. By drawing no more than 30 ma through RL, we are insuring that the current through the zener never drops below Izk. So we see that the zener diode can be used as a voltage regulator, because it keeps the voltage across itself constant with changes in input voltage, as well as for changes in load current.

Designing A Zener Regulator

Now that we know how the zener operates as a voltage regulator, all we need to know is how to calculate the value of series resistor Rs and we can design a regulated supply. Suppose we need an output voltage (E2) of 10 volts, and the maximum value of current that the load will draw is 45 ma and the minimum value is 20 ma. The value of series resistor Rs must be low enough so there is sufficient current through it when the unregulated input voltage (E1) is at its lowest value to supply the maximum load current plus the required zener current.

Say that E1 can vary between 12 and 15 volts. When E1 is 12 volts and E2 is 10 volts, the current through Rs must be the

45 ma load current plus the required minimum current for the zener. The actual value of Izk will depend on the specific zener diode used, but we can't choose a diode until we know how much power the zener will have to dissipate. So we assume a value of Izk of 5 ma, since this is typical of 1-watt and 10-watt zeners, although it may be less for smaller wattage zeners. Assuming Izk is 5 ma, and knowing the maximum load current is 45 ma, the current through series resistor Rs will be 45 + 5 or 50 ma to insure good regulation. The minimum input voltage (E1) is 12 volts, so the minimum voltage across Rs will be 12 - 10 or 2 volts. Then, using Ohm's Law, we find the value of Rs to be R equals E divided by 1 or 2 divided by 0.050 or 40 ohms. Rs cannot be larger than 40 ohms; if it is, the current through the zener drops below Izk when the load current is maximum.

We now determine the necessary wattage rating of the zener, and in order to do this we must find the maximum current through the zener. Maximum zener current occurs when input voltage E1 is at its highest value and the load current is at its lowest value. Since Rs is 40 ohms and the maximum voltage across it is 5 volts:

$$I = E/R = .125 \text{ amps or } 125 \text{ ma.}$$

Now, the maximum load current is 45 ma, so the maximum current through the zener is 125 ma less 45 ma or 80 ma. The zener has a resistance of:

$$R = E / I = 10/0.08 = 125 \text{ ohms}$$

and wattage or power equals:

$$P = I^2R = 0.0064 \times 125 = 0.81 \text{ watts}$$

and the 1 watt size would therefore suffice.

Effect of Temperature

Zener diodes are available at almost any breakdown voltage from a very few volts up to several hundred. Zener diodes, like resistors, vary somewhat from their stated values. They are available at breakdown voltage tolerances of 5, 10, and 20 percent. For example, the actual breakdown voltage of a 20-volt unit with a 10 percent tolerance might be anywhere between 18 volts and 22 volts.

Like all semiconductor materials, except tunnel diodes, the characteristics of zener diodes vary considerably with temperature. For voltages above 6 volts, the breakdown voltage increases with temperature (positive temperature coefficient). For zeners with breakdown voltages below 5 volts, the breakdown voltage decreases with temperature

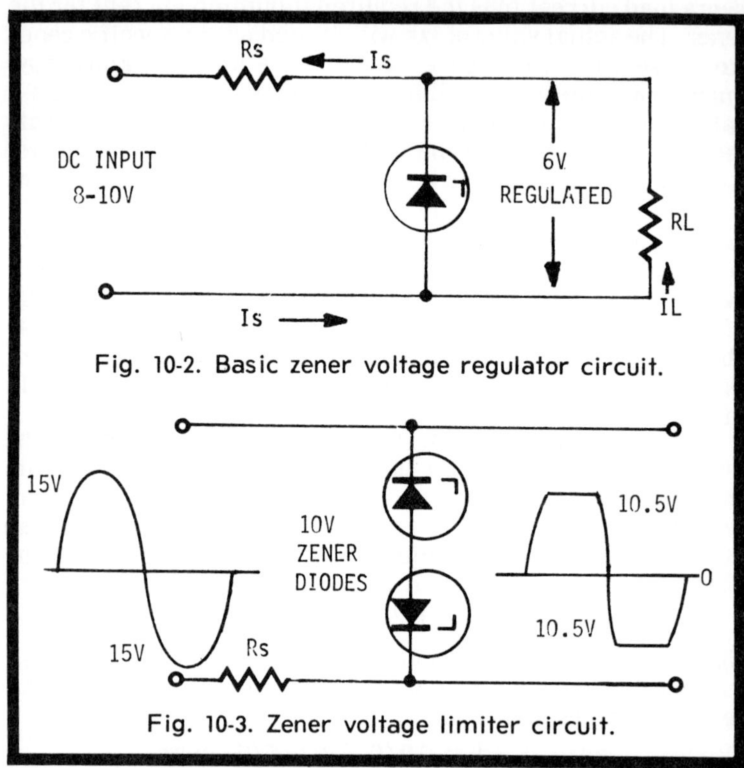

Fig. 10-2. Basic zener voltage regulator circuit.

Fig. 10-3. Zener voltage limiter circuit.

increases (negative temperature coefficient). For breakdown voltages between 5 and 6 volts, the temperature coefficient may be positive, negative, or zero, depending upon the reverse current strength. However, it is small in any case.

Since the temperature coefficient of zeners with breakdown voltages above 10 volts is high, regulation that is much less temperature-sensitive can be obtained by connecting a suitable number of 5-volt zeners in series. For example, six 5-volt zeners in series would have a breakdown voltage of 30 volts with almost a zero temperature coefficient, while the voltage breakdown point of a 30-volt zener unit will increase about 3 volts with a 20 degree F change in temperature.

Stabistor

The stabistor diode provides voltage regulation at very low voltages, and although very similar to the zener diode, it operates in the forward direction instead of reverse. Where voltages of less than 3 volts are required, the stabistor is forward biased and has a very high resistance in the forward direction until breakover is reached. At this point, conduction occurs due to the very low resistance of the device. By shunting meter movements with two stabistors in parallel, one in each direction, overload protection is assured regardless of probe polarity.

Other Uses for Zener Diodes

A zener diode is an excellent grid bias voltage regulator because it has a constant voltage drop independent of current. The conventional cathode bias resistor and bypass capacitor are replaced by a zener diode to provide the desired voltage. The zener is biased to breakdown by the plate supply voltage and the plate current through the zener diode should be somewhere between Izk and Izm. The bypass capacitor is not required because the zener itself offers a low impedance to the signal. The advantage of using the zener is that the bias voltage remains a constant voltage for any value of plate current.

A reliable voltage divider is another use for the zener and by using several units connected in series across the supply, we have regulated the output voltage at each step. Zener diodes may also be used to limit peak voltage in AC circuits by connecting them back-to-back. No matter what the polarity of the voltage across the two diodes in series, one of them will be forward-biased with a voltage drop of approximately 0.5 volt and the other reverse biased with a voltage drop equal to the voltage rating of the zeners, which is 10 volts in Fig. 10-3.

OTHER TYPE DIODES AND USES

The capacity across the barrier of the diode junction is put to use in a class of diodes known as varactor diodes. In such diodes the capacitance is not a constant value but varies with the voltage. Thus the varactor is a variable capacitor in which the value is adjusted by varying the bias voltage. The varactor finds use in automatic frequency control systems, frequency modulation, sweep generators, remote tuning control, and frequency multiplying circuits. It is also used in a class of amplifying and frequency-converting devices known as parametric amplifiers. Parametric amplifiers are microwave amplifiers having as their basic element, a reactance that can be varied periodically by an AC pump voltage.

Photodiodes

Light (like heat) can supply enough energy to a semiconductor to break covalent bonds. Light can cause three electrical effects when it shines on a light-sensitive device:

a. Photovoltaic effect: When light falls on a rectifying contact or junction, an EMF is generated across the terminals of the device. Commercial photovoltaic cells are also called photocells.

b. Photoconductive effect: In this case, light shining on a semiconductor device reduces its resistance. The energy of the light breaks covalent bonds, making additional charge carriers available. The outward effect is a reduction in resistance. So, by varying the light the resistance is varied, thereby varying the output current.

c. Photoemissive effect: When light falls on a photoemissive material, electrons are emitted from the surface of the illuminated material. Emission from a cathode of a vacuum tube is the most common commercial use of this effect. Semiconductor devices employing the photoemissive effect are not commercially available.

Junction diodes exhibit both photovoltaic and photoconductive properties. The apparent difference between cells used for their photoconductive or photovoltaic characteristics is that photoconductive cells are used with a bias battery while photocells are not. Photocells do not need batteries since the cells themselves act like batteries when exposed to light. Photocells find application in photometric equipment and light-measuring devices such as photographic light meters and colorimeters (a device used for determining color or measuring the intensity of color). Industrial applications are found in photoelectric relays, counters, and control devices. Photocells using the sun as a light source are called solar cells. Solar cells, using silicon as the semiconductor, are used almost exclusively as voltage supplies for satellites.

Thermistors

Thermistors are thermally-sensitive resistors. They are devices in which the resistance varies widely with changes in temperature. They can be made from a variety of P-type semiconductors. One common type is made from manganese, nickel, and cobalt oxides mixed in accurate proportions. Some thermistors are so temperature sensitive that their resistance doubles with a temperature change of as little as 25 degrees F. In a typical type the resistance will change by 5 percent for each degree centigrade change in temperature.

Most thermistors have a negative temperature coefficient; that is, as the temperature rises the resistance goes

234

down. However, thermistors with a positive temperature coefficient, sometimes called sensistors, are also available. With sensitors the resistance increases when the temperature increases. Aside from the many uses found in control circuits, the thermistor is very important for use in transistorized circuits where temperature causes collector current to change excessively. The effect of such changes can be minimized by a series-connected thermistor in a critical transistor lead.

Varistors

Varistors are variable resistors, and they differ from thermistors in that their resistance varies with the applied voltage rather than with temperature changes. When the voltage across a varistor is increased, its resistance decreases. Varistors were first used in lightning arrestors. When the high voltage of a lightning discharge hit the varistor, its resistance dropped to a low value so that it passed the high current of the lightning bolt. Varistors have since found applications in the general field of surge suppression and voltage stabilization. They are manufactured by pressing some suitable semiconductor material with a ceramic binder under high pressure and temperature. With respect to polarity, it does not matter which way a varistor is connected in the circuit, as its characteristics are independent of the polarity of the applied voltage.

DIODE CAPACITANCE

Just as a capacitor consists of two charged plates held close together but insulated from each other, a diode is two plates (the N-wafer and the P-wafer) close together with a high resistance between them in the reverse-current direction. Consequently, the diode also acts as a capacitor, but since the wafers are very small, the capacity is very small, typically 5 to 10 pf.

At low frequencies the capacitance effect of the diode is negligible. The higher the frequency, the easier the current passes through a capacitor. Consequently, when diodes are used in circuits operating above a few hundred kHz, the reverse current increases because the diode has a low reactance at high frequencies. If the frequency is high enough, the diode acts almost like a short circuit, and conducts freely in either direction, making it useless as a rectifier. The useful high-frequency limit of diodes can be increased by special design that reduces the capacity value. High-frequency computer diodes have a capacity of only 1 or 2 pf, making

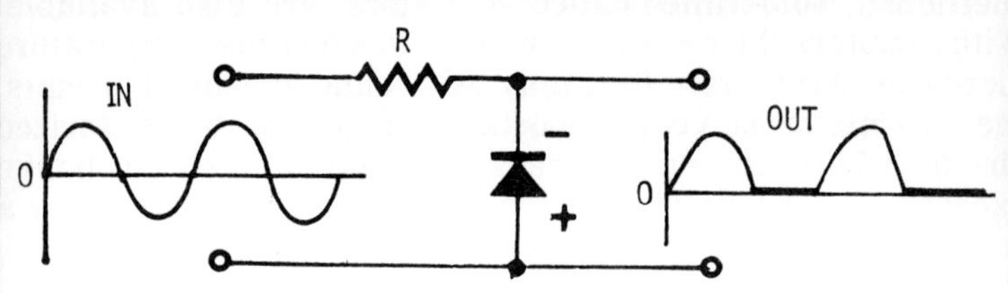

Fig. 10-4. Shunt clipper network and waveforms.

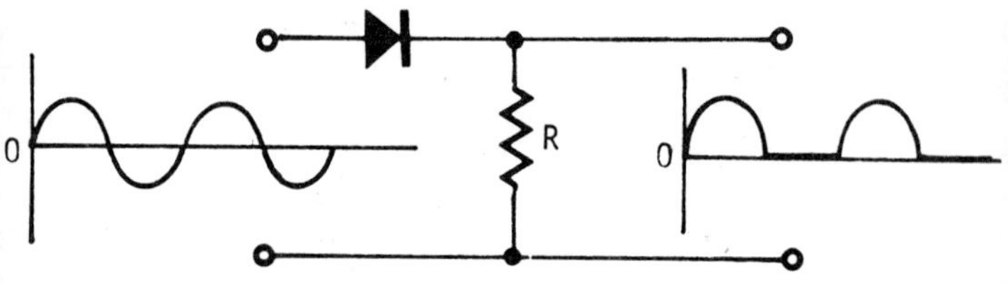

Fig. 10-5. Series clipper network and waveforms.

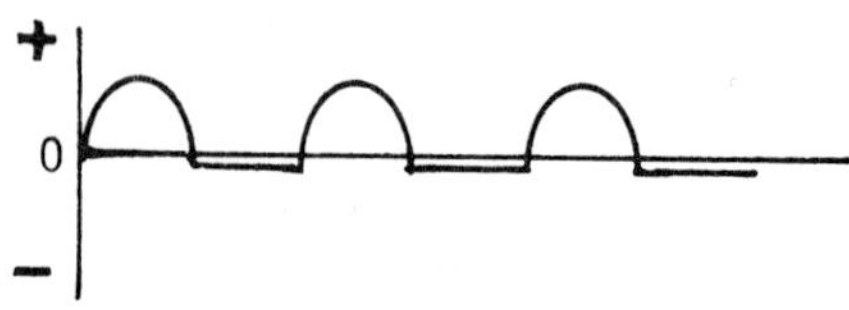

Fig. 10-6. Practical clipper output waveform.

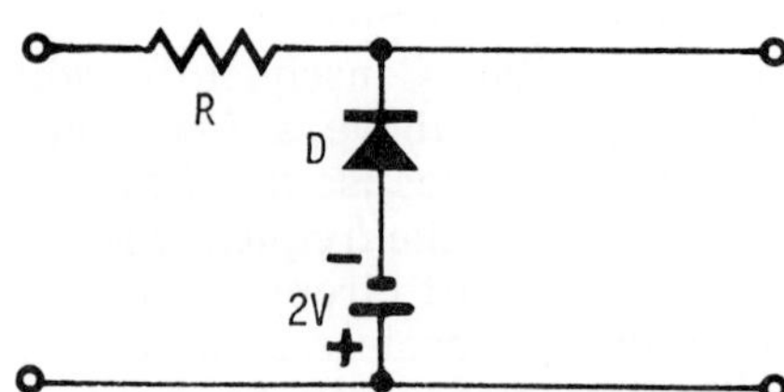

Fig. 10-7. Biased shunt clipper network.

them useful at frequencies as high as several MHz. The manufacturer's data sheets usually list the capacity of each diode type.

DIODE CLIPPING AND CLAMPING

The switching action of diodes can be used to clip off or remove unwanted portions of voltage waveforms. Such circuits are called clipping or limiting circuits, and probably the simplest of these is the half-wave rectifier, shown in Fig. 10-4. For the positive half-cycle of the input waveform, the diode is reverse-biased and acts as an open circuit. The input voltage then appears at the output terminals through resistor R. During the negative half-cycle, the diode is forward-biased and conducts, becoming a short circuit across the output. With the output a short circuit, all of the input voltage appears across resistor R and the output voltage is zero. In this way the circuit clips off all portions of the input voltage which are negative. The circuit in Fig. 10-4 is called a shunt clipper because the diode is shunted across the output and achieves its clipping action by shorting the output terminals.

Fig. 10-5 shows a series clipping circuit, where the diode is placed in series with the input and resistor R is across the output. When the input voltage is positive, the diode is forward-biased and becomes a short. The input voltage is then connected directly to the output terminals and appears there. When the input goes negative, the diode opens and the input is disconnected from the circuit. The output voltage is then zero, and in this way the negative portions of the input voltage are clipped.

Since actual diodes are not perfect, the output from a practical shunt clipper circuit will not be the ideal output as shown in Fig. 10-4. For one thing, the voltage across a conducting diode is not zero, but a few tenths of a volt. Hence, the clipping occurs slightly below 0v, as shown in Fig. 10-6. Furthermore, a practical diode doesn't change instantly from the conducting state to the nonconducting state when the applied voltage changes polarity. As a forward-biasing voltage reduces to nearly zero, the diode resistance starts to increase, so that the voltage across the diode increases. Similarly, when the voltage reverses and starts to unblock a reverse-biased diode, the diode resistance does not assume its minimum value until the forward-biasing voltage has risen to several tenths of a volt beyond the point where conduction starts. As a result, the sharp corners shown in Fig. 10-4 at the line appear rounded, as in Fig. 10-6. Sharp corners are not possible on any practical pulse.

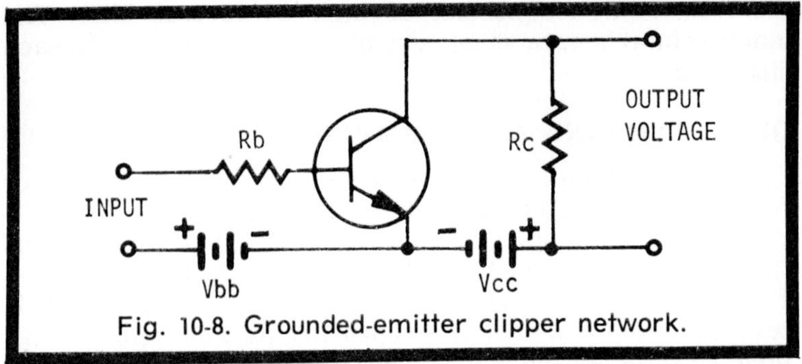

Fig. 10-8. Grounded-emitter clipper network.

Using the series clipper in Fig. 10-5, the point of clipping from a practical connection will not be 0v as shown; there is bound to be some leakage current through the diode when it is reverse-biased. Since this leakage current flows through R, there is a small voltage drop across R and, as a result, at the output, even though the diode is reverse-biased. The sharp corners are rounded just as with the shunt clipper arrangement because the switching time between conducting and non-conducting is never instantaneous.

Clipping Circuits With Biasing

Some clipping circuits may remove all of either portion of the input wave (positive or negative); there is a way to clip all signal above zero volts or all below zero volts. In fact, by adding a battery for biasing, a clipping circuit may be used to strip a portion of the wave above or below any specific value desired. The shunt clipper in Fig. 10-7 is biased by the addition of a 2v battery in series with the diode. All of the input wave more negative than - 2v will be removed regardless of the original value. The output waveform will simulate the input other than the portion below -2v which has been removed. After all, the diode acts as an open or a closed switch, and when it is acting as an open switch, the diode (and battery) play no part in the circuit operation. Therefore, during the interval in which the diode is not conducting, the output voltage is exactly equal to the input, just as though the battery and diode were not included.

When acting as a closed switch, the diode connects the battery across the output voltage and the output voltage must equal the battery voltage. The output waveform is the same as the input wave during that portion of the cycle when the diode does not conduct (open). The output waveform has a value of -2v during the portion of the cycle when the diode conducts (closed).

The output waveform of the biased shunt clipper may be established by determining when the diode conducts and when it does not; or, in other words, when it is forward-biased and when it is reverse-biased. If the input voltage is greater than the battery, it determines the polarity of the circuit and forward-biases the diode. The diode in Fig. 10-7 conducts as the input signal becomes more negative than the bias battery, as it is then forward-biased. With the input series aiding, the diode is still reverse-biased; it acts as an open switch and the output offers the exact voltage of the input. As the input wave reverses polarity, it overcomes or bucks the bias battery. When it reaches the level of the battery, the input voltage begins forward-biasing the diode, thus forcing it to conduct. This connects the bias battery across the output and pegs the voltage at that of the battery, ignoring the input completely.

TRANSISTOR CLIPPERS

Transistors act as clippers or limiters when overdriven. Overdriving the input results in cut off, while overdriving the output simply saturates the amplifier. A basic grounded- or common-emitter amplifier with an NPN transistor is shown in Fig. 10-8. With Ein negative to ground, the transistor is reverse-biased while the voltage is greater than Vbb, cutting the transistor off. As a result, the resistance from collector to emitter is quite high and the output voltage approaches zero. As the base becomes forward-biased, with the signal voltage swinging the other way, the transistor conducts and Vcc is connected across the output with the negative voltage flowing through the transistor. The output voltage may only approach, but never exceed the supply voltage.

Input voltage Ein plus battery voltage Vbb controls the base current, and when Ein exceeds -Vbb, the transistor conducts, since its base-emitter junction is forward-biased. The output follows the input, until the input signal exceeds -Vbb on its negative excursions, bucking the bias battery completely and causing the transistor to be cut off as its base-emitter junction is reverse-biased. When the transistor is cut off by negative input voltage peaks, the output becomes zero, resulting in the clipping of that output wave.

If the input voltage increase is accompanied by an increase in bias voltage Vbb to prevent cut off, clipping does not occur as before, but the positive peaks may be too great for the amplifier to handle. Since the output may never be greater than -Vcc, by overdriving the input the output tries to exceed -Vcc and in so doing, saturates. Any of the output wave more negative than the battery voltage is cut off as a result.

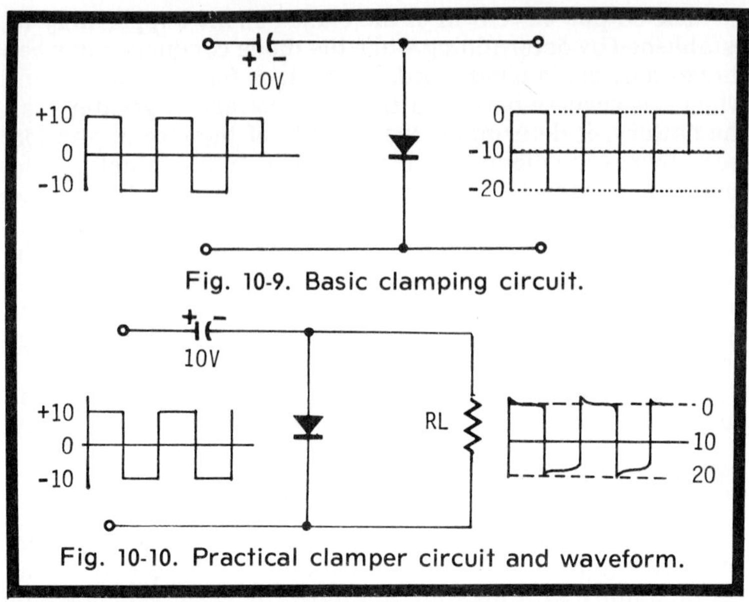

Fig. 10-9. Basic clamping circuit.

Fig. 10-10. Practical clamper circuit and waveform.

So, there are two methods of effective transistor clipping, either by cutting it off with too much in, or by saturating it with too much out; if both conditions are present simultaneously top and bottom clipping result. If this action is carried far enough, a square-wave output may be closely approximated.

CLAMPING CIRCUITS

A clamper circuit freezes the wave peak, positive, negative, or both, at a predetermined level. Without changing the actual waveshape, the addition of the clamping circuit in Fig. 10-9, provides a square-wave output of 20v peak-to-peak, the same as the input. However, the input was +10v to -10v, and the output after clamping is 0v to -20v. Since the input is symmetrical about zero, it has no DC component. On the other hand, the output is symmetrical about -10v, the average level over at least a complete cycle, which indicates a DC component.

While biasing the diode for conduction, the capacitor charges to the voltage of the battery. After charging, the polarity of the supply is reversed, and current will not flow due to the reverse-biasing on the diode. The output voltage in this reverse-biased state is equivalent to the sum of the supply and capacitor voltages.

Since the input voltage is +10v, the diode is forward-biased and acts as a short circuit across the output, thus the output must be 0v under this condition. Now with the input reversed to -10v, the diode is reverse-biased, and acts as an open circuit, permitting the output to reach -20v by adding the supply (input) voltage and series aiding capacitor voltage.

If the input voltage is continuously reversed at a regular rate between +10v and -10v, the output is a square-wave varying between 0v and -20v. This illustrates the clamping action of the circuit, and by adding a DC component it often proves useful in switching circuitry. Regularly utilized in television receivers, the clamping circuit is known also as a DC restorer because of the function it performs. Needless to say, the input wave does not necessarily need to be a square-wave; but, regardless of its shape, the input and output waves will bear the same shape and amplitude. The zero point must shift in the output so that it is the most positive point.

Assuming that the load is not great, the capacitor in the clamping circuit seldom has time to discharge to any appreciable degree. But on each positive input swing, the capacitor is recharged to the peak value. However, should the load be heavy enough to provide substantial discharge of the capacitor between peaks, the output will be distorted and will not parallel the input exactly. If significant discharge does occur, a larger value capacitor will provide the greater time constant needed to avoid the problem.

In a practical circuit Fig. 10-10 exhibits the decreased pulse amplitude present during the negative duration. The positive top is not an exact zero value, since there is a voltage drop across the diode. The voltage drop peaks near the leading edge because of the heavy rush of current to recharge the capacitor at that specific time. The time constant of the capacitor is somewhat critical for certain clamping applications. If it is too short, low-frequency distortion may be experienced. However, the time constant should not be made too long by using too large a capacitor, because several cycles could be required for the capacitor to discharge to the proper level for clamping action.

NANOAMPERE SENSING CIRCUIT

The circuit shown in Fig. 10-11 may be used as a sensitive current detector or as a voltage detector having a high input impedance. R1 is set so that the voltage at Point A is ½ to ¾ volts below the level that triggers the 2N494C. A small input current (Iin) of only 40 nanoamperes will change C2 and raise the voltage at the emitter to the triggering level. When the 2N494C is triggered, both capacitors C1 and C2 are discharged through the 27-ohm resistor which generates a positive pulse

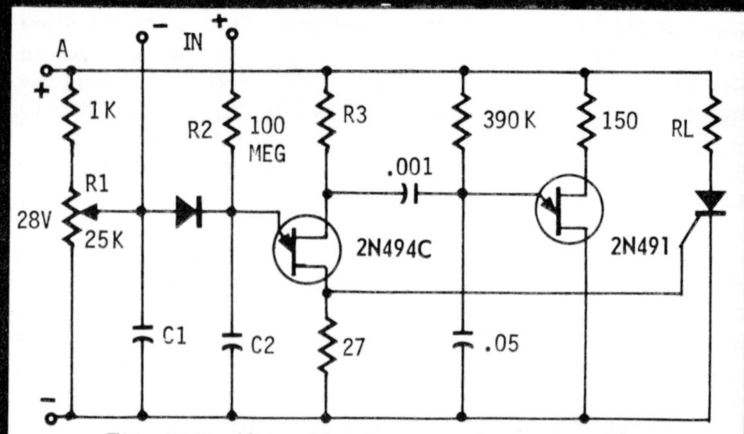

Fig. 10-11. Nanoampere current-sensing circuit.

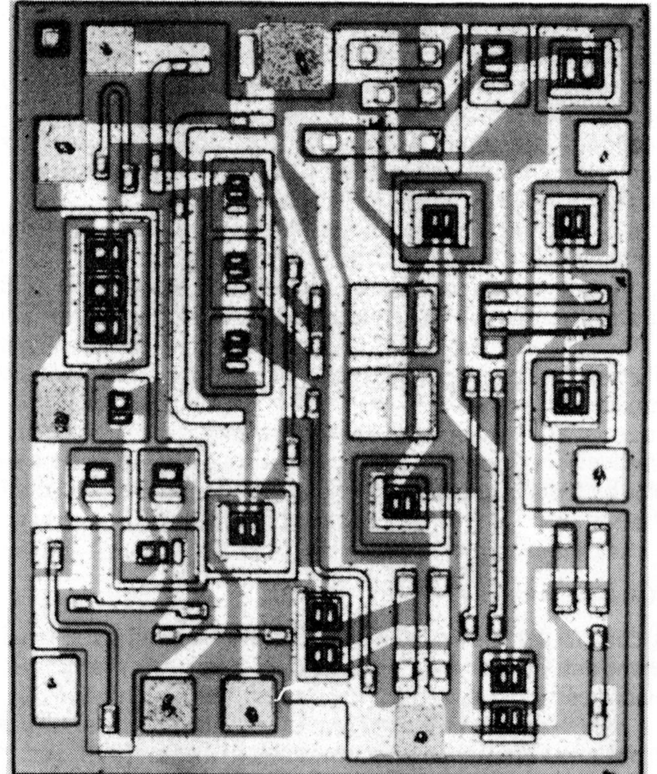

Fig. 10-12. Monolithic sense amplifier; MC 1540, MC 1440. (Courtesy Motorola Semiconductor Products)

with sufficient amplitude to trigger a silicon controlled rectifier (SCR), or other pulse-sensitive circuitry. C2 is kept small for faster triggering response time and C1 is used to provide the pulse output energy. Rapid recovery is obtained after the 2N494C triggers, since both capacitors are charged through R1.

An advantage of this circuit is that the leakage current of the silicon diode effectively subtracts from the leakage current of the unijunction and thus provides some temperature compensation. The input current available (Iin) through the 100-megohm resistor will be much lower than the minimum trigger requirement for the 2N494C (peak point current, Ip, is 2.0 ua). By pulsing the upper base of the 2N494C with a 0.75-volt negative pulse, the peak point voltage (Vp) will drop slightly, and if the voltage level at C2 is greater than this, the unijunction will trigger with the necessary Ip supplied from C2. By using this technique, the 2N494C has been triggered with external input currents (Iin) as low as 1 nanoampere with a 2000-megohm resistor for R2.

The period of the 2N491 relaxation oscillator is not critical, but it should have a time constant of 0.02, or less than that of the 2N494C. For this sensing circuit a floating power supply using a zener diode will permit grounding one of the sensing input terminals if this is desirable. R1 should be adjusted so the circuit will not trigger at the maximum ambient temperature in the absence of the current- or voltage-sensing signal. R3 can be adjusted for the best stabilization of Vp over the required temperature range. Fig. 10-12 shows the layout of an integrated circuit sense amplifier.

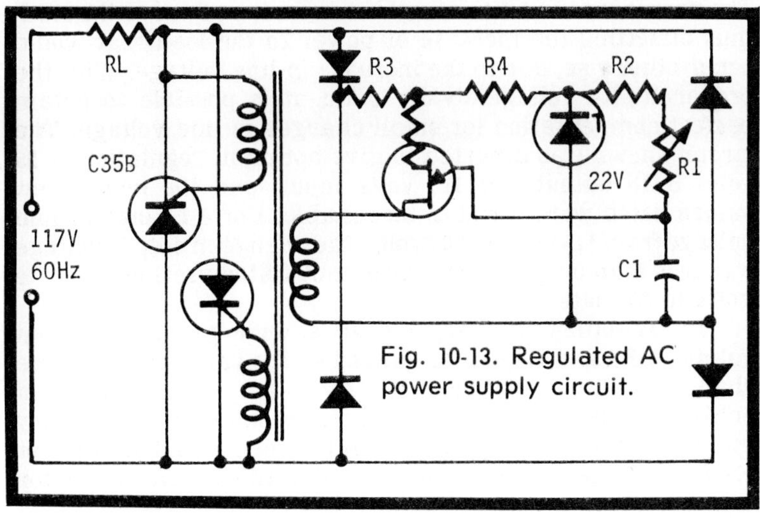

Fig. 10-13. Regulated AC power supply circuit.

A regulated AC power supply is shown in Fig. 10-13. The voltage developed across the SCRs during blocking is used for the unijunction interbase supply, as well as for synchronization of the unijunction transistor. The firing circuit is connected to the output of a single-phase bridge formed by the 1N1695s. Through the action of the bridge, the zener diode, and the resistor R3, a clipped and rectified voltage is applied to the UJT and its emitter circuit. The regulating ability of the supply is the result of R4. The capacitor charging voltage is equal to the zener voltage and is essentially constant over the half cycle prior to the instant when the SCR is triggered. The UJT interbase voltage, Vbb, is not constant during this interval, but is equal to the breakdown voltage of the zener diode plus a small fraction of the line voltage determined by the voltage-dividing ratio of R3 and R4. Thus at any given phase angle the interbase voltage will increase if the line voltage increases.

The peak point voltage of the unijunction transistor is equal to the interbase voltage times the standoff ratio and is given to correspond to the two interbase voltage curves. The emitter voltage follows the normal exponential charging characteristic since the charging voltage, V1, is constant. The UJT and SCRs trigger when the emitter voltage equals the peak point voltage.

It is readily apparent that as the line voltage increases, there is an increase in the delay before the UJT and SCRs are triggered; hence, the conduction angle of the SCRs decreases. The decreased conduction angle reduces the power to the load, thus offsetting the increase of power to the load that would occur otherwise, due to the increase in line voltage. With the proper R3-R4 voltage-divider ratio, it is possible to obtain perfect compensation for small changes in line voltage. The circuit shown was adjusted to give optimum regulation at 25 volts RMS output and 115 volts input, and the component values listed were found to be suitable. For a change in line voltage from 115 volts to 100 volts, the change in output voltage was less than 0.1 volt with any output voltage setting from 10 volts to 30 volts.

If regulation is desired over a wide range of output voltages, R1 and R4 can be ganged so the value of R4 can be changed with the position of R1. The value of R2 is chosen to achieve the desired temperature compensation in the ordinary manner. There is some interaction between the adjustment of R4 for voltage compensation and the adjustment of R2 for

244

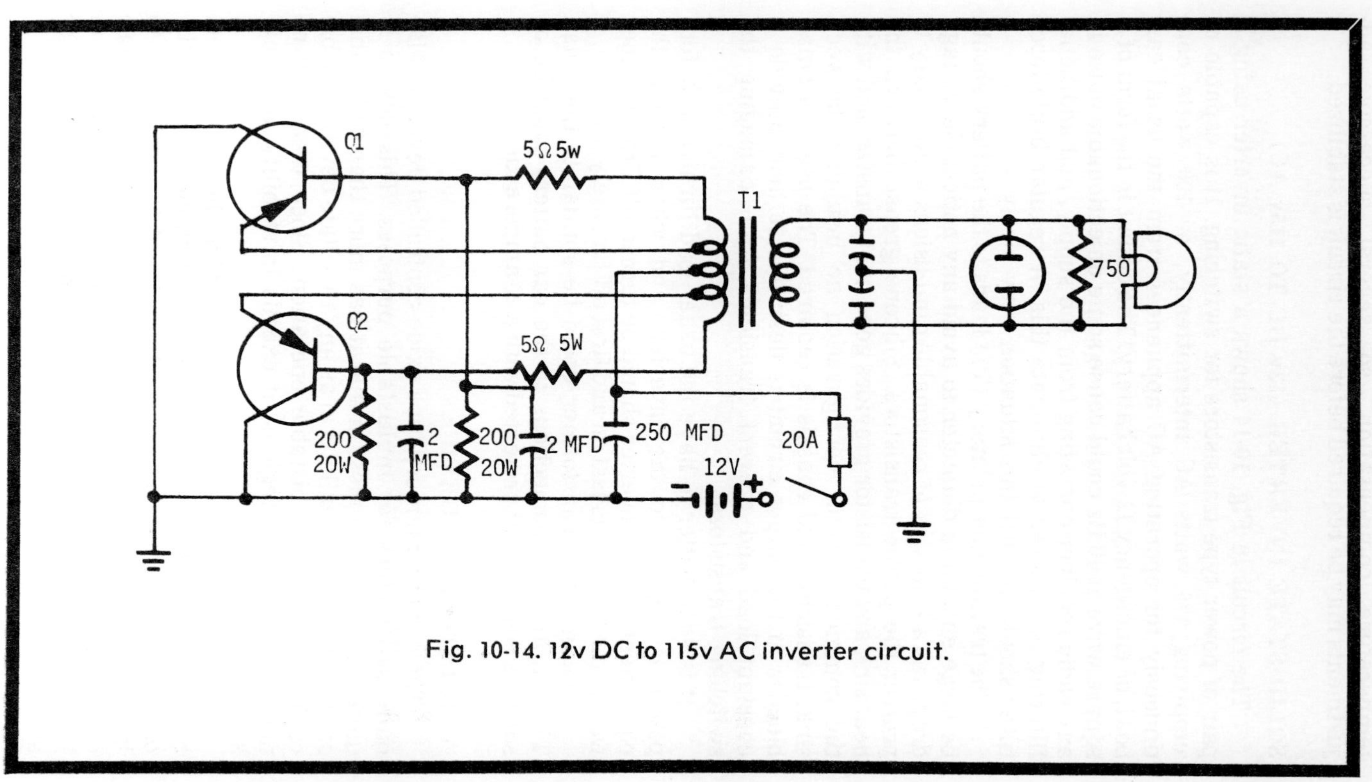

Fig. 10-14. 12v DC to 115v AC inverter circuit.

temperature compensation, so several successive adjustments may be required before the circuit is stabilized.

SOLID-STATE INVERTER (12v DC TO 115v AC)

The circuit in Fig. 10-14 shows a static inverter using a pair of power type transistors for switching. It is capable of supplying 175 watts AC intermittently, or 100 watts continuously for operating AC appliances from the usual car, boat, or emergency 12-volt battery. The output in the form of a square wave at 60 Hz could cause some objectionable noise in any radio receiver operating from the supply, and additional filtering is required to eliminate this. A regular brute-force filter should correct this situation very nicely.

The transformer primary (T1) leads to the battery should be large enough in diameter to avoid any noticeable voltage drop, and a heat sink of reasonable dimensions is necessary to protect the power transistors. Silicone grease between the heat sink and transistor provides good heat transfer, and with the common-collector configuration no insulation between sink, transistor, and chassis is required. The use of a small bleeder in the output circuit is desirable to help provide a constant load and prevent transients from damaging the switching transistors.

It is even better to have the usual load turned on before power is applied to the inverter, and when turning the equipment off, the inverter should be turned off first. In other words, the inverter is last on and first off for safety's sake.

The inverter transformer could be similar to the Triad TY-75A, and the connections to the car battery must be as short as possible and protected with a 20 ampere fuse.

A MULTISTABLE DEVICE

The use of a memory-type oxide-controlled device is also quite satisfactory for multistable purposes. This device is quite different from most in that it has more than two stable states, and yet is as unilateral as any two-state device capable of functioning in a bistable operation. Since there are no negative-resistance regions, it excells in stability and transient considerations.

Following the operation of the device, beginning in its high-voltage (low-current) state 1, a pulse applied between the emitter and oxide-layer causes the gain of the transistor to increase; then, it rests in state 2 (less voltage and more current than 1). By applying another pulse of like polarity to base and oxide-layer, the transistor moves into state 3 or saturation.

In order to return it to its original state 1, it is necessary to apply a pulse of opposite polarity to the base and oxide-layer terminals. The driving source must be a high impedance to assure that sufficient charge is stored in the oxide-layer each time the pulse is applied. The time constant of the source must be long, and by careful adjustment of the oxide thickness, the time constant can exceed 90 minutes.

INTEGRATED CIRCUIT MAINTENANCE

Servicing and maintaining IC equipment requires different techniques. Ordinarily, the technician would have complete access to the various components of any circuit composed of tubes, semiconductors, capacitors, resistors, transformers, coils, etc. The integrated circuit, by its very construction, defies repair or replacement of its internal

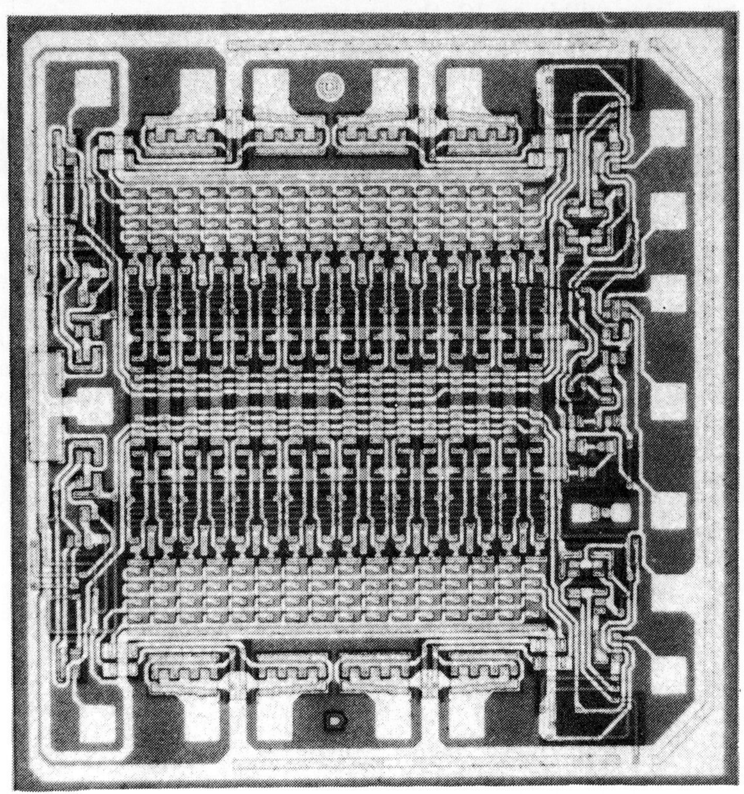

Fig. 10-15. 128-bit read-only memory; XC 170, XC 171. (Courtesy Motorola Semiconductor Products)

elements without rendering the package useless, at least for normal use. Since the circuit elements (resistor, transistors, diodes) form a definite part of the silicon substrate, they are, therefore, not detachable, and the hermiticity of the package would be destroyed by any attempted replacement of elements. This leaves only one course open, that of replacing the entire defective unit.

There may be quite a large number of units involved, so the first move is to reduce this number. However, before dividing the troubles to smaller segments, a quick glance at certain necessary precautionary measures may be well taken. The circuit elements are minute, and the power levels are likewise very small. It is necessary to be aware of the voltage and current limitations involved in the various circuits. A particular unit could easily be damaged by the voltage from an ohmmeter used for testing purposes. The polarity is also very important; by simply applying the reverse polarity to a transistor within an IC, the junction could be destroyed permanently. As mentioned before, special care is always mandatory when dealing with the ever popular field-effect transistors of the metal oxide type (MOSFETs); **the gate must be grounded when the device is not connected to the circuit.** It would otherwise be normal for stray charges to accumulate on the oxide layer and render the device completely worthless.

All operating power to integrated circuits should be regulated properly and be entirely void of transients. The internal impedance of the supply must be low and possible points of regenerative feedback eliminated. A complete knowledge of operating conditions, currents, voltages, signal type, and level of the integrated circuits is needed in order to isolate the defective device or devices. The use of test points so frequently provided and reference to manufacturer's test and performance data always proves worthwhile.

Chapter 11
Miscellaneous Switching Circuits

Universal time constant curves enable us to find current or voltage at any time after a step voltage is applied. The curves, as shown in Fig. 3-11 (Chapter 3), can be used with RC as well as RL circuits. For example, we may wish to find the output voltage in Fig. 11-1 at 0.5s after closing the switch. Since one time constant is 0.25s, 0.5s is equal to two time constants. We have a rising curve in Fig. 11-1, so we use the rising curve on the time constant chart. Checking the chart, we see that at 2 time constants, the voltage reaches 87 percent of its steady-state value. The steady-state value in Fig. 11-1 is 32v. Hence, the voltage after 0.5s following closing of the switch is 0.87 times 32 or 27.8v. The current at this same moment would be 0.87 times 2 or 1.74A.

In order to find the output voltage in Fig. 11-2 at 0.5s after the switch is closed, we use the bottom curve of the universal time constant (Fig. 3-11) because of the decaying voltage involved. We find that at 2 time constants, the voltage is 13 percent of its maximum value. Thus 0.13 times 32 equals 4.16v.

When a step voltage is applied to an RC circuit, the time required for the voltage across the resistor to fall to 37 percent of its maximum value is called one time constant. This length of time in an RC circuit is calculated from the formula:

$$t = C \times R$$

where t equals time constant in seconds, R is the resistance in ohms, C the capacity in farads; or of more practical use, R in megohms and C in mocrofarads; t is the time constant in seconds. At the same time, the voltage across the capacitor has reached 63 percent of its maximum value.

A similar situation exists when resistance and inductance are in series. Considering the inductance L to have no resistance, and with R at zero, closing the switch would send a rapidly changing current through the circuit. Back EMF

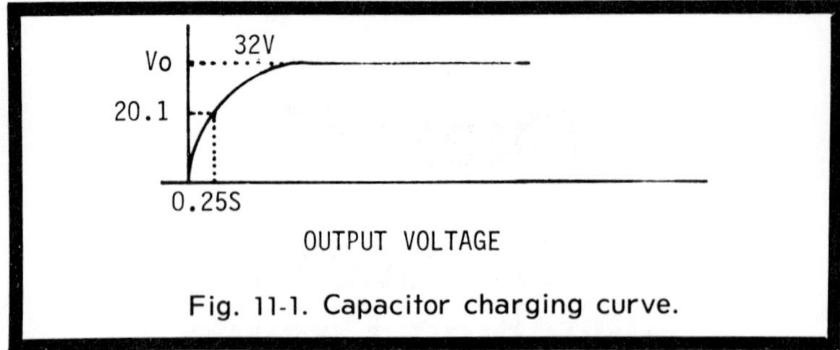

Fig. 11-1. Capacitor charging curve.

developed by the self inductance of L would be almost equal but opposite to the applied EMF. The result is that the initial current is very small, and the back voltage depends upon the change in current; the back voltage would cease to offer opposition if the current did not continue to increase.

The time constant of an inductive circuit is the time in seconds required for the current to reach 63 percent of its final value, and equals L in henrys divided by the resistance in ohms,

or: $t = L/R$

The resistance in the wire of a coil acts as though it is in series with the inductor.

An inductor cannot be discharged in the same way as a capacitor, since the magnetic field disappears as soon as current flow ceases. The energy stored in the magnetic field instantly returns to the circuit when the switch is opened. The rapid disappearance of the field causes a very large voltage to be induced in the coil (many times the applied voltage), because the induced voltage is proportional to the speed with which the field changes. The usual result when opening the switch in any circuit with inductance alone is that a spark or arc forms at the switch contacts at the instant it is opened. If the inductance is large, and the current in the circuit is high, a great deal of energy is released in a very short period of time. Under such conditions, the switch contacts may burn or melt. The spark across the opening switch or relay contacts may be suppressed by connecting a resistor and capacitor in series across the contacts.

Time constants play an important part in many devices such as electronic switches, timing, and control devices as well as pulse-forming and shaping units. In nearly all applications, RC time constants are involved, and often it is necessary to know the voltage across the capacitor at some time interval larger or smaller than the actual time constant

250

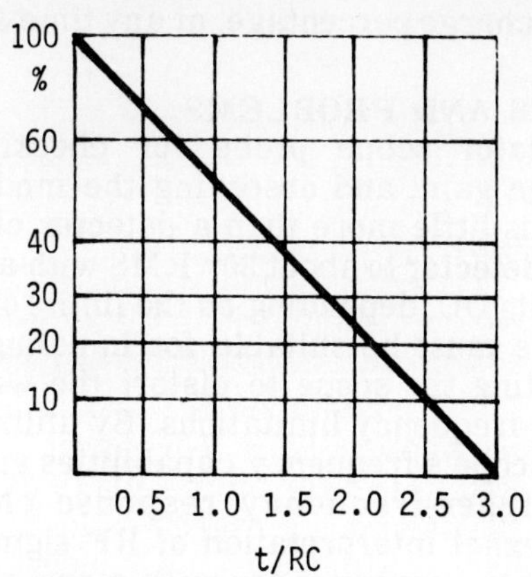

Fig. 11-2. Graph showing voltages across a capacitor during discharge.

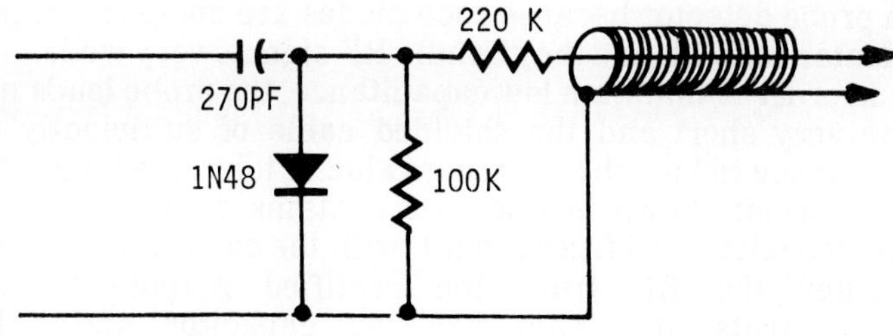

Fig. 11-3. Demodulator probe circuit.

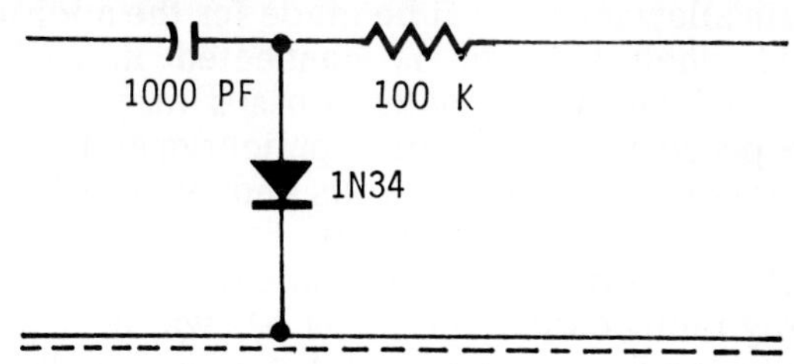

Fig. 11-4. High-impedance VTVM probe circuit.

of the circuit as given by the formula. From the graph in Fig. 11-2 you can determine the voltage across a capacitor, in terms of initial charge percentage, at any time after discharge starts.

PROBE TYPES AND PROBLEMS

A demodulator scope probe for checking distortion, measuring stage gain, and observing the modulation of RF and IF signals is little more than a detector circuit. It functions as an AM detector to about 30v RMS with a tip maximum of 500 to 1000 volts DC, depending on the limits of the capacitor used. The probe must be suitable for handling such signals without permitting the scope to distort the waveshape as a result of its low-frequency limitations. By utilizing the proper RF probe, the scope's frequency capabilities are expanded to include the higher-frequency response characteristics necessary for exact interpretation of RF signals.

The necessary properties for such a probe are low shunt capacitance in order to avoid detuning or loading the circuit, and a detector to convert the high-frequency signal to video or audio if the RF is modulated or to DC if the signal to be observed is pure RF. A germanium or silicon diode is often used as a probe detector because such diodes are compact, require no heater current, and they handle RF signals very well.

In order to maintain low capacitance, the probe leads need to be very short and the shielded cable of sufficiently low capacitance to hold the shunting value to tolerable levels. The probe circuit shown in Fig. 11-3 contains a diode detector, input capacitor, and filter circuit, with the cable capacity used to filter the RF from the rectified output. In such arrangements, the input blocking capacitor and output isolating resistor avoid circuit disturbances during signal measurements. By coupling a signal generator to the grid or base input circuit of one stage and connecting the demodulator probe to the input of the next stage, the gain of an IF or video stage can readily be measured on the scope.

Certain allowances must be made for the additional losses introduced when a probe is connected in series with a capacitor and the signal point. Losses may amount to appreciable percentages in some applications. Test procedure simplifications are quite numerous and must be improvised to properly fit the situation at hand.

An RF probe using a metal shield and designed for use with a high-impedance meter is shown in Fig. 11-4. By multiplying the DC reading by 1.4, the peak value of the RF wave is indicated. The RF voltage range is determined by the peak inverse rating of the diode. The PIV usually ranges from 40 to 50v and may be extended by connecting diodes in series.

252

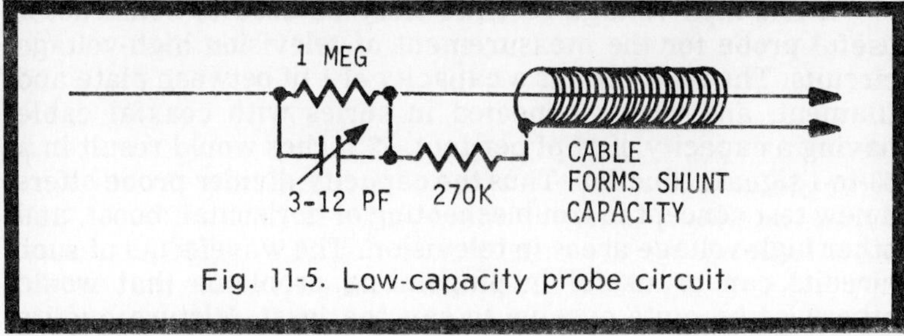

Fig. 11-5. Low-capacity probe circuit.

Low-Capacity Probe

Circuit loading and signal distortion are minimized by a low-capacitance probe, and the flexible shielded cable feeding the input circuits of the test equipment reduces the possibility of external interference with the desired signal. However, the shielded lead adds considerable capacitance across or in shunt with the circuit being measured. The very means of protection against one problem is instrumental in creating another.

Adding this input capacitance to the scope's vertical circuit, may amount to 75 pf, which could produce intolerable distortion of the present pulses in video or sweep circuits. The higher the circuit impedance, the greater the effect of the capacity. A low-capacity probe corrects the error with a capacitor and resistor as pictured in Fig. 11-5; the resistors maintain the low-frequency attenuation and the capacitors the high-frequency signal reduction. The trimmer must be adjusted to the required value for a linear response curve. Too much capacity causes an upward bend or irregularity of the trace, while insufficient capacity results in a downward display. Although a probe will attenuate a signal to a marked degree, the vertical gain of the scope will more than make up for the loss. Response to 25 MHz is normal and the average DC limit is 600v.

High-Voltage DC Probe

A high-voltage probe usually has a multiplier factor of 100 on the DC ranges of 11-megohm input meters, providing an 1100-megohm input impedance for measuring high-voltage circuits to 30 KV. High-voltage measurement by capacitance voltage division is quite useful as a result of its lack of circuit loading characteristics. The voltage across series capacitors in high-voltage circuits is inversely proportional to the capacity. The greater the capacity, the lower the voltage drop since resistance is lower.

A 1X2 high-voltage rectifier may be used to construct a useful probe for the measurement of television high-voltage circuits. The rectifier has a capacity of 1 pf between plate and filament, and when connected in series with coaxial cable having a capacity of 12 pf per foot, 49 inches would result in a 50-to-1 signal reduction. Thus the capacity-divider probe offers a new test concept in troubleshooting of horizontal, boost, and other high-voltage areas in television. The waveforms of such circuits can be useful in pinpointing problems that would otherwise be quite evasive to say the least. Picture quality could be impaired considerably by such simple problems as a lower rise time or amplitude of the horizontal pulse. Any discrepancy in the pulse rise time or amplitude is immediately apparent with the use of the capacity-divider probe.

By slightly reducing the high-voltage, the picture is dimmer and requires a higher setting of the brightness control to compensate. This increases the deflection senstivity, making the picture larger, but the boost has decreased and lowered the gain of the deflection amplifiers. Thus the sensitivity increase is balanced out by the decrease in deflection gain, but the end result is a sizable loss in overall picture texture and quality.

SCOPE QUALITY

The laboratory or professional type oscilloscope may prove extremely useful in the design and measurement of pulse circuitry. The time calibration of the sweep enables the user to measure rise time, phase angle, and decay time with ease and amazing accuracy. A slow sweep frequency of less than 1 Hz is made possible by a triggered sweep, and the cathode followers in series with the feedback paths are instrumental in providing the high-frequency response so essential to the accurate measurement of pulse rise times. Improved linearity also results from the extras in the sweep circuits and makes the higher cost of a quality instrument a bargain in most cases. The weight of these considerations can hardly be over-emphasized in view of the true value of the conclusions made as ε result.

254

Index